Help us make the

Here at Make:, we share stories and projects and create celebrations called Maker Faires because we believe in one simple idea:

We are all makers.

We've brought you Make: magazine since 2005, and more than 1 million people participate in Maker Faires every year across 44 countries.

Together we've built a movement and a global community — but there's so much more to do!

Join us in recognizing the value of the global maker movement and its impact on many, many lives. Your participation will bring the community closer together, help us share more projects and knowledge, and develop new ways to learn.

Become a member today and, together, we can make more makers tomorrow.

Dale Dougherty
Founder of **Make:**

Become a **Make:** member
Learn about the benefits of membership at make.co

CONTENTS

Make: **Volume 63** June/July 2018

ON THE COVER:
Lisa and Mike Winter face off in a robot battle of machine vs. human control

Photo: Hep Svadja

Hair and Makeup Rachel Lusk rachellusk.com

Font: Desyrel by Apostrophic Laboratories

Get Smart

Hep Svadja, Sophy Wong, Geoff Decker – Hidden Vision Photography, Glen Scott, Shing Yin Kohr, @EspressoBuzz

40 **46**

60 **72**
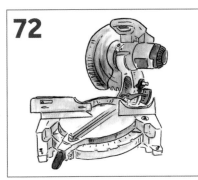

14

22

Make:

EXECUTIVE
CHAIRMAN & CEO
Dale Dougherty
dale@makermedia.com

CFO & COO
Todd Sotkiewicz
todd@makermedia.com

EDITORIAL

EDITORIAL DIRECTOR
Roger Stewart
roger@makermedia.com

EXECUTIVE EDITOR
Mike Senese
mike@makermedia.com

SENIOR EDITORS
Keith Hammond
khammond@makermedia.com

Caleb Kraft
caleb@makermedia.com

EDITOR
Laurie Barton

PRODUCTION MANAGER
Craig Couden

BOOKS EDITOR
Patrick Di Justo

EDITORIAL INTERN
Jordan Ramée

CONTRIBUTING EDITORS
William Gurstelle
Charles Platt
Matt Stultz

CONTRIBUTING WRITERS
Carlos Campos, Jeremy S. Cook,
Marc de Vinck, DC Denison,
Kelly Egan, Trey German, Heidi
Hinkel, John Iovine, Ben Krasnow,
Melissa Lamoreaux, Gordon
McComb, Forrest M. Mims III,
Carol Reiley, Glen Scott, Micah
Scott, Angela Sheehan, Greg
Voronin, Lindy Wilkins, Mike
Winter, Sophy Wong

DESIGN, PHOTOGRAPHY & VIDEO

ART DIRECTOR
Juliann Brown

PHOTO EDITOR
Hep Svadja

SENIOR VIDEO PRODUCER
Tyler Winegarner

MAKEZINE.COM

ENGINEERING MANAGER
Jazmine Livingston

WEB/PRODUCT
DEVELOPMENT
Rio Roth-Barreiro
Pravisti Shrestha
Stephanie Stokes
Alicia Williams

CONTRIBUTING ARTISTS
Ryan Garcia, Shing Yin Khor,
vecteezy.com

ONLINE CONTRIBUTORS
Gareth Branwyn, Chiara
Cechini, Jon Christian, Gretchen
Giles, Liam Grace-Flood,
Sarah Hijazi, Darrell Maloney,
Goli Mohammadi, Andrew
Morgan, Dan Royer, Daniel
Schneiderman, Evan Sheline,
Katelyn Sheline, Greg Treseder,
Steven White

PARTNERSHIPS & ADVERTISING

makermedia.com/
contact-sales or
partnerships@
makezine.com

DIRECTOR OF
PARTNERSHIPS &
PROGRAMS
Katie D. Kunde

STRATEGIC
PARTNERSHIPS
Cecily Benzon
Brigitte Mullin

DIRECTOR OF MEDIA
OPERATIONS
Mara Lincoln

DIGITAL PRODUCT STRATEGY

SENIOR DIRECTOR,
CONSUMER
EXPERIENCE
Clair Whitmer

MAKER FAIRE

MANAGING DIRECTOR
Sabrina Merlo

MAKER SHARE

DIGITAL COMMUNITY
PRODUCT MANAGER
Matthew A. Dalton

COMMERCE

PRODUCT MARKETING
MANAGER
Ian Wang

OPERATIONS MANAGER
Rob Bullington

PUBLISHED BY

MAKER MEDIA, INC.
Dale Dougherty

Copyright © 2018
Maker Media, Inc.
All rights reserved.
Reproduction without
permission is prohibited.
Printed in the USA by
Schumann Printers, Inc.

Comments may
be sent to:
editor@makezine.com

Visit us online:
makezine.com

Follow us:
🐦 @make @makerfaire
@makershed
google.com/+make
f makemagazine
makemagazine
makemagazine
twitch.tv/make
makemagazine

Manage your account
online, including change
of address:
makezine.com/account
866-289-8847 toll-free
in U.S. and Canada
818-487-2037,
5 a.m.–5 p.m., PST
cs@readerservices.
makezine.com

ASK THE EDITORS

What's your favorite board and why?

Mike Senese
Executive Editor

Arduino Esplora. Now discontinued, it was ahead of its time with a built-in joystick and array of controller buttons — plus sound, light, and temperature sensors.

Hep Svadja
Photo Editor

Circuit Playground Express! Circuit Python compatible, M0 processor, tons of sensors and goodies onboard, and MakeCode friendly means easy programming for the nervous novice!

Caleb Kraft
Senior Editor

Adafruit Trinket because it is tiny, powerful, cheap, and can do HID, which means I can use it to emulate keyboards and mice for people with physical disabilities.

Matt Stultz
Contributing Editor

Arduino Uno. With the greatest level of community support along with compatible shields, libraries, and example code, nothing beats it.

Juliann Brown
Art Director

It used to be a skateboard. Then it was a snowboard. Now it's a cutting board.

Keith Hammond
Senior Editor

My much-dinged, oft-patched 6'10" Gordon & Smith speed egg, shaped by Chris Darby: fast and cruisey but still snappy off the top. Also covet the amazing 9'0" Pearson Arrow Formula 1!

Issue No. 63, June/July 2018. *Make:* (ISSN 1556-2336) is published bimonthly by Maker Media, Inc. in the months of January, March, May, July, September, and November. Maker Media is located at 1700 Montgomery Street, Suite 240, San Francisco, CA 94111. SUBSCRIPTIONS: Send all subscription requests to *Make:*, P.O. Box 17046, North Hollywood, CA 91615-9588 or subscribe online at makezine.com/offer or via phone at (866) 289-8847 (U.S. and Canada); all other countries call (818) 487-2037. Subscriptions are available for $34.99 for 1 year (6 issues) in the United States; in Canada: $39.99 USD; all other countries: $50.09 USD. Periodicals Postage Paid at San Francisco, CA, and at additional mailing offices. POSTMASTER: Send address changes to *Make:*, P.O. Box 17046, North Hollywood, CA 91615-9588. Canada Post Publications Mail Agreement Number 41129568. CANADA POSTMASTER: Send address changes to: Maker Media, PO Box 456, Niagara Falls, ON L2E 6V2

PRINTED WITH
SOY INK

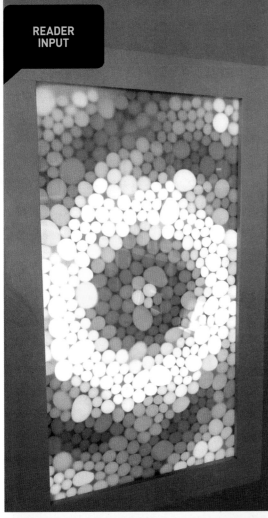

Making It Your Own

THANKS FOR KEEPING *MAKE:* GREAT

I've been a longtime subscriber and just wanted to thank you for keeping the ideas fresh and varied. I have made countless projects inspired by *Make:*, but my current project is one of my favorites. Inspired by the last issue, I made a "Rainbow Lightbox" (Volume 61, page 46) around a 24" monitor running continuous Mandelbrot set videos sourced from a Raspberry Pi. An Alexa-enabled outlet plug allows me to turn it on and off hands-free, as well as to have it come on automatically on a schedule. This was a great *Make:* project utilizing skills I have learned from years reading *Make:* — woodworking, painting, acrylic cutting, as well as simple Raspberry Pi and Alexa programming. Keep the great ideas coming! –*Russell Radoff, M.D., via email*

INSPIRED INNOVATION

What I like about the magazine is that it inspires me to make other projects using some of the hardware and software ideas I'm now introduced to through these articles. I'm currently working on a tide clock that was inspired by the traffic monitor article ("Travel Light," Volume 59, page 52). Standard tide clocks only work if your tide is 6 hours 13 minutes long. They rarely are, so my project calls the WorldTides API to achieve the required accuracy. I would never have done that without the lead from that article using the Google maps API.

–*Mark Williams, via email*

Prison Ban of the Month

» **LOCATION:** California Dept. of Corrections and Rehabilitation

» **TITLE:** *Make:* Vol. 62

» **REASON:** Plans to disrupt the order, or breach the security, of any facility. Your publication contained material on page(s) 52–65.

» *EDITOR'S NOTE: Of all the articles in the spy issue, we didn't think the Dragonfly Rubber Band Helicopter would be one to cause trouble!*

Russell Radoff, Mark Williams

Adrian Chadd
March 18 at 10:19pm

I wasn't going to re-subscribe to Make (because well, no time) but then Nora grabbed the latest copy and started asking who the girl on the cover was and questions about all the tech.

So, I just resubscribed to Make Magazine again.

👍 Like   💬 Comment

DIY AI

BY DALE DOUGHERTY, founder and CEO of *Make:*

On a recent episode of *Science Friday*, Ira Flatow asked former astronaut Sandra Magnus what it was like to live for 4½ months in the International Space Station. She replied that, "life up there is really magical." She explained that living in microgravity gives you a much greater appreciation for it and what excited her was that more and more people would soon be getting access. Then she added: "There is a difference between intellectual knowledge and experiential knowledge." She said you can learn about microgravity as a phenomena but "when you live in microgravity you understand it in a different way." Magnus said "getting people up there who have technical training and creative sparks to actually internalize what that microgravity environment is all about will really expand the possibilities of what we may be doing in space ..."

THE DISTINCTION BETWEEN knowledge and experience turns out to be important in artificial intelligence. The first generation of artificial intelligence in the '70s set out to represent knowledge symbolically such that computers could make decisions as humans do. Researchers were soon overwhelmed by the complexity of representing all knowledge and the processing requirements overwhelmed the computers of the day.

More recent progress in artificial intelligence has come in part from the growth in speed and power of networked computers. However, unlike some kind of general artificial intelligence, AI today is more like tools (in the physical sense) or libraries (in the software sense) that are trained to do specific things well. They are trained by data to learn and get feedback to improve their decision-making powers. In the case of visual processing, the AI tool or library might distinguish a cat from a dog — without really knowing what a cat or a dog is. It is trained to identify a pattern of pixels in a set of images, which are labeled cat or dog. As it looks at unlabeled images, using its past training to decide, its mistakes are fed back to it so that it can learn from them. That is why the term "machine learning" is used to signify this new generation of AI.

Machine learning is based on experiential learning, the same kind of learning that we recognize in makers. The machine is not being told what to do by a program but rather it is figuring out what to do and getting better at it. The goal of machine learning is to exceed what humans can do — even exceeding what its programmer told it to do.

There is a lot of money going into developing AI. Companies have loads of data that can be used to train these systems and they have large numbers of users that will provide the necessary feedback. Nonetheless, I hope that AI tools do not become the exclusive domain of large companies and well-funded startups. Makers can democratize AI while others capitalize on it. There's a need for DIY AI and for makers to explore a wide range of applications. (Fortunately, there are a number of open source AI libraries.) In "How to Make an AI Robot," Mike Winter provides instructions and encouragement on getting started (page 22). Greg Voronin shows you how to use Mycroft AI to create an interactive rochambeau game (page 28). What is possible now might seem silly or insignificant. Yet makers can use AI to solve problems that may never get addressed. The same technology that threatens to replace people and take away jobs can empower others to do things that only an elite few have been able to do in the past. ◗

HELLO_

vecteezy.com

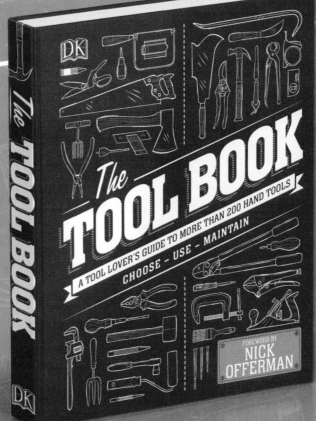
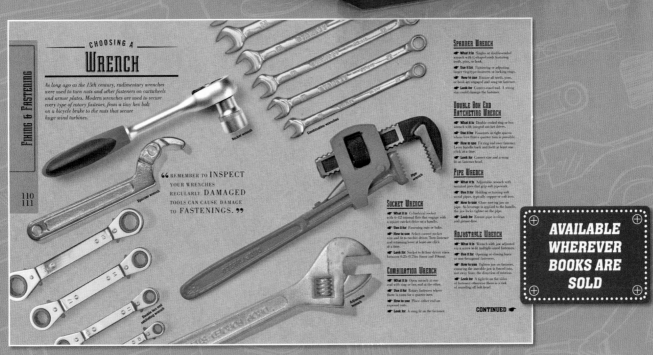

MADE ON EARTH

Backyard builds from around the globe

Know a project that would be perfect for Made on Earth?
Let us know: *makezine.com/contribute*

DIVINE DISCARDS

GABRIELDISHAW.COM

Though he uses castaway circuit boards and other thrown-out electronics for his art, **Gabriel Dishaw's** sculptures are anything but junk. Each piece incorporates its own complex pattern of colors, with many presenting a character from geek culture with a vibrant flamboyancy.

"I truly found my passion for this particular art form in 9th grade," Dishaw says. At the time, Dishaw's instructor let class members choose from among 30 different ideas for an upcoming art project. One of the suggestions, junk art, stuck with Dishaw, who went home to research the topic. After trying it out, he fell in love with the style and he's been perfecting his craft ever since.

"Generally, I begin a project without any defined plan," Dishaw says when describing his process. "Instead, I look at how well pieces work together or inspire a theme and then let those materials drive the color scheme or overall endpoint of where an idea will land, what it will turn into. For instance, the inspiration for one of my horse sculptures, *Rearing Horse*, came to me while taking apart an old adding machine. Some of the pieces reminded me of a horse's head. The rest just fell into place."

Dishaw went on to say his smaller projects take him approximately 40 hours to complete, plus however long it takes to scavenge for the hundreds of parts each sculpture needs. As he puts it, the success of each sculpture relies as much on luck as it does skill. — *Jordan Ramée*

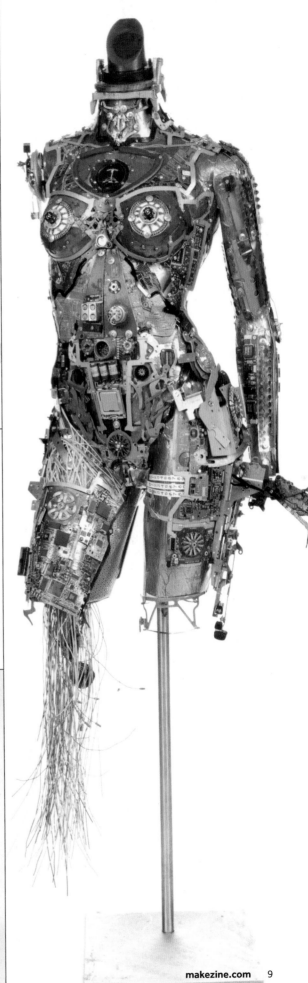

PIXEL PAINTING

SPRAYPRINTER.COM

When it comes to large format art, there's no tool more versatile than the spray paint can — it's fairly cheap, massively portable, and the paint sticks to nearly any surface. But when your natural design medium is more iPad than New York subway train, how do you get your vision out on the streets? Enter SprayPrinter.

Inventor **Mihkel Joala** came up with the idea after his daughter asked him to paint a mural of a unicorn on her bedroom wall. Short on artistic talent, he had to improvise to get the job done. After considering the combined technologies of the Wii controller and a car's fuel injection valve, SprayPrinter was born.

SprayPrinter is a tool that snaps right onto the top of your favorite can of Krylon. A phone app maps the image you want to create onto the target wall, and the smart spray nozzle has an LED that lets the app track its position. All you need to do is get the can where it needs to go, and the custom spray valve does all the rest of the work.

There's no need to adhere to the monotony of monochrome — the design app lets you break up your clever creation into multiple color layers. Want to go bigger, or just give your arms a break? SprayPrinter can be easily adapted to a number of styles of plotter gantry systems too.

—*Tyler Winegarner*

SprayPrinter

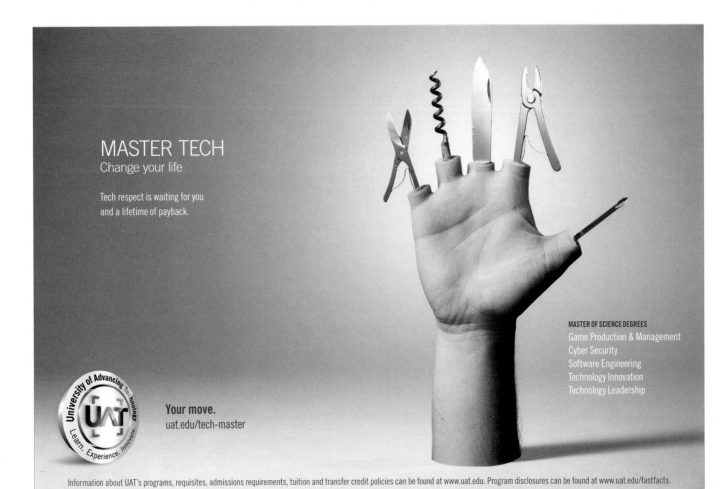

CAPTIVE CIRCUITS INSTRUCTABLES.COM/ID/LED-RESIN-CUBES

Like insects trapped in amber, these circuits evoke the feeling of time being stopped. The fragile wires and delicate mercury switch so much like the legs and body of a frail mosquito, but this circuit is very much alive, and sucks power instead of blood.

Marcus Dunn wanted to learn how to work with resin and thought it might be smart to start with something small, since resin is quite expensive. He became enamored with the look of

The entire circuit is encased in clear resin, leaving only a charging port. When you tip it, the circuit turns on and off, cycling through different color modes on the LEDs.

Though this design is quite stunning, Dunn learned some lessons and has thoughts on how to make it even better. He's considering powering it from a small solar panel instead, which would allow the entire circuit to be embedded within the block of

WHERE THERE'S SMOKE

Written by Jordan Ramée

The Flaming Lotus Girls' dazzling oversized art has been firing up crowds for 18 years

JORDAN RAMÉE spends most of his time writing about geek culture. Although he's particularly passionate about game design and Japanese art, he loves traveling around the world to meet creators from all walks of life.

STARTING AS A GROUP OF EIGHT IN 2000, THE FLAMING LOTUS GIRLS has grown into an organization of artistic acclaim over 100 members strong. Named after their first project, a giant lotus incorporating a 30-foot liquid-fuel flame effect, the Flaming Lotus Girls continue to build what their fans love: fire-spouting, metalworking marvels. Though the group is primarily composed of women, no one is turned away so long as they have a passion for making and an eagerness to learn.

A WARM WELCOME

Flaming Lotus Girls is, as the group calls it, a "do-ocracy." Caroline "mills" Miller, a member who helps design fire effects and safety measures, says, "[Our group] is twofold: if you want to build it, then you get to design it; and if you show up and stay involved, then you get to make the decisions." Although the idea of giving a voice to every member may seem like an easy way to fall prey to decision gridlocks and constant arguments, the system has worked for years. Every project begins with a brainstorming session where every member is given a chance to voice ideas they think the group should pursue, or themes or elements that would be cool to explore.

"Once we have some early ideas on the table, we might break into smaller groups based on what folks are most excited about, and begin to flesh out specifics about a proposal," Miller says. "Eventually, around three proposals come to the group for voting, and the idea with the most votes gets pushed to a final proposal submission." By the end of the process, the final idea tends to be very different from whatever was originally proposed, but everyone can be happy they contributed.

Miller goes on to explain how this do-ocracy extends beyond the planning process. "In addition to the physical aspects of the art, members also contribute and develop leadership and administrative skills, such as fundraising, PR, photography, crew care, logistics, and all the other activities needed to support a group such as ours," she says. Providing access to specific types of tools and machines, like CNC cutters, also changes from build to build, and every project requires different teams of people to transport, unload, and reassemble the pieces whenever it is going to be displayed.

Even events such as Burning Man require, as the group humorously puts it, "many, many spreadsheets." Regardless of the type of maker, all can contribute to Flaming Lotus Girls' process. They need more than welders and pyrotechnicians.

FIRED UP

The Flaming Lotus Girls' creations are all sights to behold, from *Seven Sisters*, a collective of interactive flaming sculptures inspired by the Pleiades constellation, to *The Serpent Mother* (opposite), a kinetic sculpture sprouting 40-foot flames and trailing almost 170 feet in length. Each sculpture blazes to life at the push of a button, making the group fan favorites among both adults and children at events such as Burning Man and Maker Faire.

"I will never forget the time I watched a couple get married underneath a sculpture that I built," one member says. "Or the first time I saw an FLG poofer go off and shrieked along with all the other uninitiated people. Or the time a little girl took a few deep breaths to psych herself up, took the controller from me, looked me square in the eyes, and said 'I am not afraid.' The best part about our community is that we all want the same thing — to push the button just one more time!"

FANNING THE FLAMES

Because of Flaming Lotus Girls' diverse set of projects (both in how they look and what they are able to do), no one can quite agree which build is the most impressive. Each member has their favorite, and fan feedback doesn't yield any definitive patterns either. *The Serpent Mother* has certainly proven to be the most popular with fans. Even after 12 years, demand to see the project in person hasn't diminished.

However, some members point out *Mutopia*, a sculpture that tells the story of the evolution of seedpod-like creatures, has the best display of fire effects, and *Tympani Lambada*, a combination of fire, vibrations, and sound working together to create a uniquely stimulating sensation, is the most remarkable example of structural engineering. "Some of us," Miller adds, "think the most impressive thing we've built is our community and fan base."

"The work we do," she continues, "could only be done with many enthusiastic and cooperating hands, and that rewards us with the feeling of tremendous accomplishment." Another member describes the community within their group as "a warm blanket that wraps you up and makes you feel loved, especially when you are in with both feet and working your volunteer butt off to make the seemingly impossible happen. The enormous communal effort it takes to build and show fire art is what binds us together and is a bright light in our individual lives." ◐

Tympani Lambada

"The enormous communal effort it takes to build and show fire art is what binds us together and is a bright light in our individual lives."

Pulse

Tympani Lambada

Mutopia

Tympani Lambada

Caroline Miller, Joe Dacanay, Kimberly Sikora

Android Apparatus by Little Dada (Lindy Wilkins and Hillary Predko, littledada.ca), modeled by Vanita Butrsingkorn.

The wildly varying ideas between DIY and consumer "**wearable electronics**" need to converge

AS A PROFESSOR OF "WEARABLE TECHNOLOGY," I find the term difficult to explain to the rest of the world. DIY electronics communities, focused on fashion-tech and hobbyist wearables, have diverged from consumer wearable electronics. From materials to design, their processes are fundamentally different. I think they should be more actively drawn together.

The Arduino LilyPad microcontroller board, which marked the beginning of DIY electronics taking physical forms that work well with our bodies, turned 11 this year. It's time we saw a greater convergence of the creativity and humanistic form language of DIY wearables — enabled by these boards — reflected in our consumer electronics.

DIY: Designing for Bodies
I teach wearable tech at a number of universities in Toronto — in contexts ranging from fashion to cyborgs —

CLOTHES *Minded*

Written by Lindy Wilkins

LINDY WILKINS is a cyborg, technologist, and community builder who specializes in lasers, whimsical robotics, and wearable technology.

where I guide students using an array of weird and wacky iterative design processes. We tend to focus on the single commonality we all have: a body! Everyone in the room can relate to moving around and wearing clothing.

Designing for bodies is fundamentally different from designing for screens, paper, or interactive spaces. Once you start strapping electronics to yourself, you quickly realize just how much design went into our clothing, shoes, and jewelry, that they so seamlessly integrate into our daily movements and activities.

We study conductive fabrics and how to sew a circuit. We learn that the shapes of electronic components are important because they're part of a *form language* — the types of shapes found in a particular design context. Our bodies are full of smooth edges and curves, while electronics typically have hard edges and points. Sewable electronics like the Adafruit Flora and Arduino LilyPad help bridge the gap between electronics and clothing using a form language that better fits the body, such as flatter, rounded form factors with sewable connectors.

Many of the current wearable technology tools accessible to DIYers rely on craft and handmade techniques to create "soft" electronics. Very few courses can say students learned how to spin their own (conductive) yarn, how to solder, and how to make their own sensors — our wearable tech classes blend traditional craft and cutting-edge technology, with lots of room for students to align themselves wherever they're comfortable on that continuum. The devices they create are often conceptual, abstract, and sometimes purely unique fashion statements.

Consumer: Designing for Moore's Law

Inevitably, someone asks how to "make it real": How do you take a project, made from hand-embroidered conductive textiles and round microcontrollers, and turn it into what we see on the market today?

Search for "wearable tech" on any shopping website, and you'll see watches and fitness trackers that feel very different from what we make in class. This is where I see a fracture in the culture around wearable technology. The vast majority of wearable technology sits on your wrist, because that's what we're comfortable with. Sewable components and soft electronics are largely left out.

We've made technology the same way for so long — hard circuit boards with flat panels — with great success, but our bodies won't feel comfortable with flat panels.

Commercial wearables feel more concerned about how small we can make the technology, rather than making it fit our bodies. The focus on miniaturization is made possible because of Moore's law, which describes the alarming rate at which technology is shrinking while simultaneously gaining computing power.

The Convergence?

It's only in the last 10 years that we've begun to think deliberately about the body as an interface, and to design for it with a user-centric mindset, partly because of that rapid shrinking of technology.

If we want technology to be part of our bodies, it has to feel like an extension of ourselves, and that involves considering the tech alongside the aesthetic. In commercial wearables, we often see amazingly powerful technology void of any specific wearability concepts or form language, like the Muse headband, a very interesting brain wave sensing technology that's also incredibly invasive, with no meaningful application.

Every so often, an experimental Kickstarter campaign pops up promising to change everything, but they rarely stick around. Technology moves so quickly, and our bodies are so picky. The smallest discomfort can be the downfall of a technology.

We're at an interesting point in the hype cycle of wearable tech. The curve of enthusiasm over a technology spikes at the beginning when everyone is excited about its possibilities. A myriad of inventions are created, followed by a sharp decline as we realize these so-called innovations don't live up to the hype, like Google Glass. It's only then that we're able to look objectively at technologies and make useful and widely adopted products.

Happily, we are exiting the initial phase of wearables enthusiasm. Smart watches and fitness trackers are becoming commonplace instead of a fancy gimmick. Their aesthetic form is beginning to change, respecting the rules laid out by the Wearable design community. Brands like Fitbit are diverging from traditional watch designs into sleek, rounded designs like the Flex 2 that find the balance between form language and information display.

But still, there's a divide between "hard" and "soft" electronics communities that perpetuates the problem. The materials used by experimental wearable tech enthusiasts are virtually nonexistent in the commercial world. This, unfortunately, is what happens when our electronics are made with fundamentally different mindsets. I think we can do more to stitch them together. ✔

Max Lander, Edward Ross, Lindy Wilkins

Re:Familiar (The Drone Dress) by Little Dada explores the potential for relationships with nonhuman entities. Like a witch's "familiar," a drone is servant, spy, and companion all at once. Here a Parrot AR drone follows the model and carries the dress' train, surrounding her with billowing silk chiffon.

The dress is fabricated with a body suit built from Ethernet cables covered in photoluminescent pigments, ethereal UV LEDs, and blower fans. We have reimagined the body in relationship to the server room, a key site of information exchange. Modeled by Carmen Ng.

Android Apparatus is a glowing piece of cyber armor custom-fit over a black leotard worn by an aerial hoop (lyra) performer. LEDs respond to an accelerometer built into the garment: As the dancer performs, the costume brightens, dims, and changes color with both the range and intensity of motion.

To enhance her performance without hindering her movement, we created a heat map to see where her body made contact with the hoop. The non-dominant arm and the chest don't contact the hoop, so they're ideal areas for the armor.

We also iteratively combined an accelerometer data visualization with the pattern while laser cutting and testing the form. The final piece is laser-cut vegetable-tanned leather, formed by traditional leather molding techniques.

Cardboard Control

Written by Marc de Vinck

COMBINING ANALOG MAKING WITH DIGITAL PLAY, NINTENDO LABO CELEBRATES THE JOY OF DISCOVERY

FROM SEEMINGLY OUT OF NOWHERE, THE INTERNET WAS ABUZZ THIS SPRING ABOUT NINTENDO'S LATEST OFFERING, LABO. Creating your own Nintendo Switch controllers out of simple cardboard has *Make:* written all over it, and we jumped at the chance to get hands-on with the system.

Cool Kits

The first build we attempted was the telescoping fishing rod, and yes, we were hooked. Nintendo has a long history of innovation in the gaming industry. Remember the Virtual Boy circa 1995, or the Nintendo Power Glove from 1989? Both of these products were way ahead of their time. In fact, it's been almost 30 years since they launched, and virtual reality gaming is still considered to be in an early stage of development.

What really sets the company apart from other console video game manufacturers has been their fanatical approach to giving a great user experience, and Labo proved to be no exception. A charmingly animated instructional sequence on the Switch guides you through the builds with ease. The transparency of the transition from digital gaming to analog building makes for an almost magical experience. The parts all fold and assemble beautifully, and the final product is surprisingly robust — and best of all, real!

Custom Creations

After playing around with a motorcycle and a robot demo kit, we were introduced to another element of Labo: their visual programming language called the Toy-Con Garage. It allows users to harness all the amazing sensors of the Nintendo Switch and its controller to build their own creations. Just drag and drop components around the screen to create interactive toys, instruments, and machines. With just a few programming blocks you can turn the Switch remote into a door sensor and build an alarm. Even more impressive is leveraging the object recognition of the Switch controller's onboard camera to track movement and react accordingly. Creating programs is simple, but not simplistic — a great opportunity for kids and parents to get together and practice their maker skills.

With Nintendo, you know this is just a hint of what's to come. We can't wait to see what's next for the Labo platform. ✐

MARC DE VINCK
is a professor at Lehigh University, teaching in the technical entrepreneurship master's program.

Nintendo

Tailoring Tech

MakeFashion inspires women to challenge the status quo of wearables

DC DENISON is the co-editor of the *Maker Pro Newsletter*, which covers the intersection of makers and business, and is the senior editor, technology at Acquia.

MakeFashion has been combining fashion and technology since 2012. This group has produced over 60 wearable tech garments and showcased at more than 40 international events. Recently, MakeFashion completed a successful Kickstarter fundraising campaign for StitchKit, an Arduino-based wearable technology toolkit. We caught up with Shannon and Maria Elena Hoover, who, along with Chelsea Klukas, co-founded the project.

What's the goal of MakeFashion?

Augmenting the human experience. Wearables can be about more than quantified self and learning about the environment. Wearable technology can be about making your experience more real and in the moment. You don't have to stop, take a phone out of your pocket, and start tapping on it with your finger.

Wearable technology is about showing who you are and telling your story. Unfortunately, tech companies have mostly ignored the fashion aspect of wearables. That's a little bit bewildering to us. Three trillion dollars a year is spent on fashion and how we look. It's obviously important to people.

How is your StitchKit product different from the many wearables offered by companies like Adafruit?

We love Adafruit's products, but Adafruit is targeting makers and hobbyists — people who are comfortable with a soldering iron. We're dealing with people who are coming into this tech world terrified. They have no previous experience. So we're focused on lowering the barriers.

I read that the MakeFashion community is mostly female. Turns out fashion is a very savvy way to get young women interested in technology.

Absolutely. If you want to get young women interested in science and technology, it's much easier to start with something they are already passionate about. If we can make it easier for these girls to innovate with fashion, they will do amazing things, and teach us all valuable lessons.

What kind of challenges are still out there, for wearable technology?

One of the big problems right now is that most technology is not meant to be worn. It's meant to be encased in a plastic or aluminum case. So durability and safety are still big issues that we're dealing with.

What groups will push the boundaries of wearable technology and fashion in the next few years?

Specialized industries that are creating wearables to fit specific needs are where the innovation will come from. We're already seeing very creative things come out of the Burning Man community. Cosplayers are also driving innovation. ◉

[+] See more at makefashion.ca

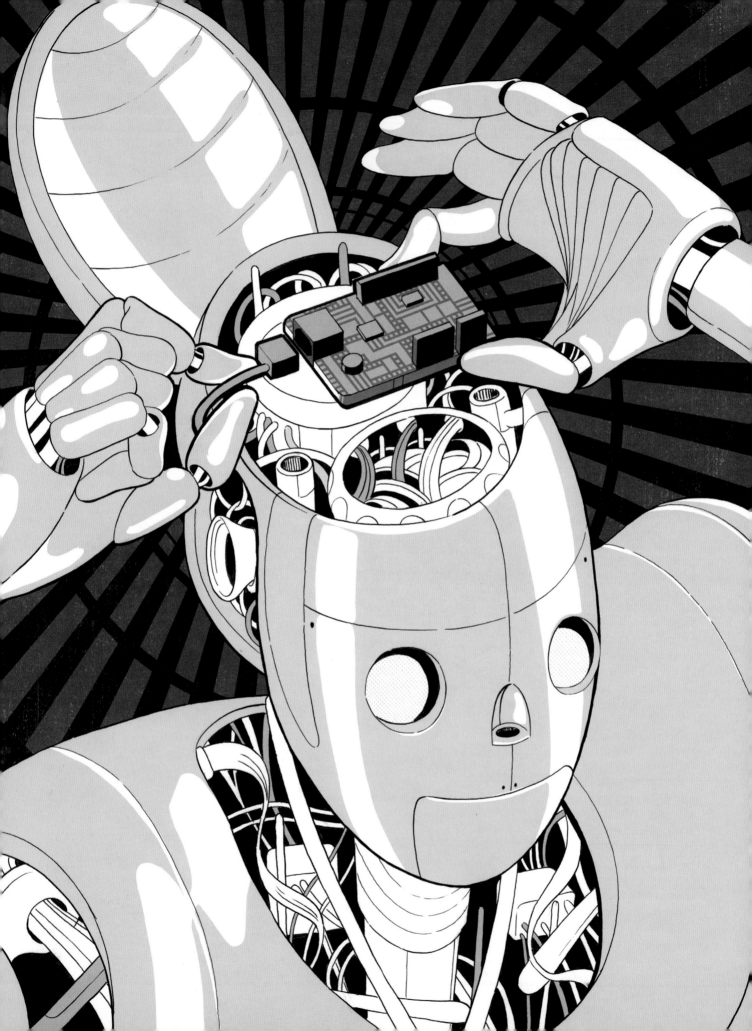

Get Smart

FIRE UP YOUR BOARDS AND PLUNGE INTO ARTIFICIAL INTELLIGENCE

Written by Carol Reiley
Illustrated by Ryan Garcia

MAKE: HAS ALWAYS EMPOWERED INDIVIDUALS TO EXPLORE HOW WE CAN get our robots to sense and feel, and examine the ways to help them think.

In this issue, we look at the latest and greatest tools from the electronics prototyping community, from large companies like Texas Instruments to maker-focused suppliers like Adafruit and SparkFun. Our Board Guide will help you find the perfect microcontroller for your project — robot or not.

We'll also delve into how these boards are being used to help push the AI world forward. Artificial intelligence will power the next generation of inventions, and we want to make it accessible to everyone, not just elite academics or big corporations.

STARTING OFF

Getting established in AI can be intimidating: Should you install TensorFlow, Torch, PyTorch, or Caffe programming framework? How many layers should you build your neural net?

What hardware should you set up for your deep learning infrastructure? But here's the thing — there are so many tools available and it's fairly quick to get started.

To dive more deeply and build a foundational background, check out both Prof. Andrew Ng's free online courses "Intro to Machine Learning" (coursera.org/learn/machine-learning) and Deeplearning.ai's five-course specialization series. Or go through the *Deep Learning* textbook by Yoshua Bengio, Ian Goodfellow, and Aaron Courville from MIT Press. We need more help solving the world's problems through AI.

The rule of thumb is to pick a problem that a human can solve in less than 1 second (AI is not quite there yet to tackle more complex problems). Review your linear algebra, probability, and calculus. And practice basic coding fundamentals (we recommend Python or Julia). We're here to assist you in acquiring the tools to help you solve big problems. Share with us what you build and have fun!

CAROL REILEY is a serial entrepreneur, roboticist, and recovering academic. She is a frequent contributor to *Make:* magazine and Maker Faire, and was featured on the cover of *Make:* Vol. 29 in 2011.

Carol fell in love with robotics and AI over 15 years ago, building highly regulated products in applications such as space, underwater, and medical. She has published over a dozen papers and patents and served as the youngest board member of the IEEE Robotics and Automation Society. She did her graduate work at Johns Hopkins University, worked on surgical robotics at Intuitive Surgical, and co-founded drive.ai — one of the hottest self-driving car startups in Silicon Valley which has raised over $77M. As a passion project she founded Tinkerbelle Labs, consisting of low-cost, do it yourself healthcare hacks, and authored and published a children's book called *Making a Splash!* about growth mindset and the psychology of success, which has sold over 15K copies. She is currently building a new startup.

A Mind of Its Own

Written by Mike Winter

WANT YOUR HIGH-TECH PROJECT TO GAIN **ARTIFICIAL INTELLIGENCE**? THIS OVERVIEW EXPLAINS THE STEPS.

Mike Winter has spent his life writing software, solving problems, and building robots for places like Google. Along with his daughter Lisa (page 26), he's most notable as a longtime BattleBot alum. Now he's focused on artificial intelligence — specifically determining if a computer-controlled combat robot can best a human-controlled robot opponent in an endeavor he calls "AI or Die." Always the educator, he guides the new maker through the basics of exploring AI robotics in this piece.

PERHAPS THE BIGGEST CHALLENGE FOR A MAKER IS TO CREATE SOMETHING SMARTER THAN ONESELF, like a robot with artificial intelligence. Since the 1950s there has been much developmental work on AI robotics but with very limited results. We don't see smart robots performing anything but the simplest tasks. Just ask a robot to fold a shirt. Even if it's a specialized AI shirt-folding robot, you should be prepared to wait for hours.

But things are changing swiftly as the interest in AI rapidly expands. In the San Francisco area hundreds of AI startups were founded last year, some creating self-driving cars, others developing digital personal assistants, IoT devices, and toys. On the global front this year, China announced a plan to become the world leader in AI by 2030. It has been predicted the need for human labor will be reduced to near zero in the next 20 to 50 years. Robots need to become much smarter to make the gains in productivity necessary for that kind of economic and cultural advancement. This is where the maker community comes into play. They can supply the innovation and hands-on practicality needed to make the required highly functional robots.

CREATING AN AI ROBOT CONSISTS OF:
- Building a robot that has a simple CPU
- Adding a second but more powerful computer for the AI
- Inventing an AI and running it on the powerful computer

Then you flip the "on" switch and see what happens! If it fails, rethink and try again. This will allow you to have fun experimenting with various machine intelligences that can operate on a real robot in the real world. Add a variety of sensors including webcams, drive motors, treads, and anything else you dream up.

1. MAKE A ROBOT

To start, I advise using a kit — it's a complete design, cheaper than buying parts separately, and incurs less shipping costs. The robot needs to hold two CPUs, two motors, a battery, and a few other things. A plate about 7" or 8" square should work well (Figure Ⓐ).

Select a kit with "Tank Steering." It allows robots to change course by having the treads go different speeds or directions. The robot will rotate about its center, which makes it easy for the AI to calculate turns. Almost any robot kit found online will work for your build. I've used both the Tank and the Triangular Tank robot kits from OSEPP with excellent results; you can see a portion of a robot I made with these parts in Figure Ⓑ. I love kit bashing.

> **NOTE:** Where treads are mentioned, wheels can be used. I use treads because they look more interesting. Never make a dull robot.

Next, add electronics to the robot. This is the motor controller, batteries, switch, I/O CPU, and voltage regulator that allow the robot to power on and move. I advise following the electronics recommendations of the robot kit you've selected. If they don't suggest a specific motor controller, I'd use a Sabertooth dual 5A — but you can find many other great options online. Advanced builders might try to customize and match their components; if you go that route, this guide is helpful: robotshop.com/blog/en/how-do-i-choose-a-battery-8-3585

Now the wiring. First create the power circuit. The motor controller's job is to send a percentage of the battery power to the motors as this controls the speed and direction of the robot. Connect the battery, on/off switch, and motor controller as in (Figure Ⓒ).

The voltage regulator converts the battery voltage to the correct level for the I/O CPU and other logical devices, like a sonar sensor. Wire the voltage regulator as in (Figure Ⓓ). Next we power the I/O CPU and sonar sensor. Wire them to the output of the voltage regulator (Figure Ⓔ). Now connect the logic wires of the sonar sensor and motor controller devices to the CPU's I/O pins (Figure Ⓕ).

Ⓐ **Robot base concept**

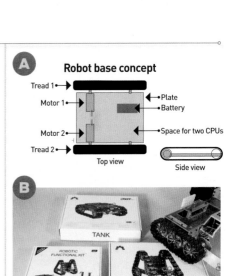

Tread 1 · Motor 1 · Motor 2 · Tread 2 · Plate · Battery · Space for two CPUs · Top view · Side view

Ⓑ

TANK · ROBOTIC FUNCTIONAL KIT · TRIANGULAR TANK

Ⓒ **Create power circuit**

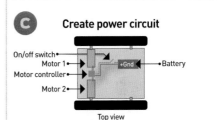

On/off switch · Motor 1 · Motor controller · Motor 2 · +Gnd · Battery · Top view

Ⓓ **Add voltage regulator**

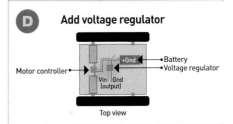

Motor controller · +Gnd · Vin Gnd (output) · Battery · Voltage regulator · Top view

Ⓔ **Power CPU and sensor**

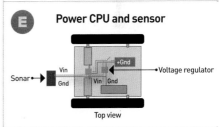

Sonar · Vin Gnd · +Gnd · Vin Gnd · Voltage regulator · Top view

Ⓕ **Connect sensor and controllers to CPU**

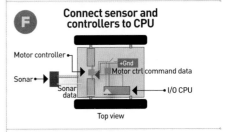

Motor controller · Sonar · +Gnd · Motor ctrl command data · Sonar data · I/O CPU · Top view

Ⓖ **AI CPU**

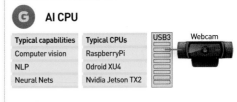

Typical capabilities	Typical CPUs
Computer vision	RaspberryPi
NLP	Odroid XU4
Neural Nets	Nvidia Jetson TX2

USB3 · Webcam

With multiple cameras and sensors, Cam1Bot can track, analyze, and record the robot contestants in the AI or Die Arena.

Whose bot is best? Mike pits his AI-controlled AIV4 against his daughter Lisa's manually driven fighter.

Hep Svadja, Mike Winter

H

2. ADD A POWERFUL COMPUTER FOR THE AI

Now that we have a basic robot built, let's dramatically increase its thinking capabilities. To do this, we'll add a new powerful computer, the AI CPU, and connect it to the existing I/O CPU. The I/O CPU will receive commands from the AI CPU, interpret them, and finally send them off to the motor controller and other devices. The I/O CPU also reads data from sensors and sends it to the AI CPU.

The AI CPU (Figure G, previous page) will run an AI model that processes video streams and other sensor data. It outputs commands to the I/O CPU. For example, if the AI CPU's computer vision detects a cat, it could send a message to the I/O CPU to turn on the motors so the robot would run away.

FACTORS IN SELECTING AN AI CPU:

- **Computer Vision:** Does the CPU have the horsepower to run it? If not the robot will need to use other sensors (like lidar) to understand its environment. The OpenCV computer vision library is free and easy to use. You will need a Raspberry Pi CPU or something faster. I use the Odroid XU4 with its eight Heterogeneous Multi-Processing (HMP) cores, because many of my robots use two webcams and have two copies of OpenCV computer vision running simultaneously. For seriously fast computer vision check out CPUs that have a graphics processing unit (GPU) and support for OpenCV (like the Jetson TX series).

- **Size:** Make sure the AI CPU will fit on your robot. I mention this because a number of people just bungee a laptop onto their robot. For some kits you will need to add a larger base plate, which can easily be made with some aluminum or wood from a hardware store.

- **Logic Levels:** This is the voltage that the CPU's digital I/O pins see as a "1."

Ground is a "0". Some CPU's pins need 5V, other 3.3V, and the Odroid XU4 is 1.8V. The I/O pins must match the digital device voltage. So if you get a motor controller that requires 5V then the CPU should output 5V. If you have a voltage mismatch there are level converter boards that go between 3.3V and 5V.

- **USB:** The faster the better. I would not use anything slower than USB3.
- **Preinstalled OpenCV** and other AI tools are a big plus

SUGGESTED AI CPUS:

- **Raspberry Pi**, the crowd favorite
- **Odroid XU4**, my favorite with its 8 cores and USB3
- **Intel i7**, use a standalone headless i7 or just mount a laptop to a (big) robot base
- **Nvidia Jetson TX1 or TX2**, super powerful, can do neural nets, but has a steep learning curve for the developer

Figure **H** shows all the parts for making an entire AI robot. Foreground shows an Arduino compatible I/O CPU (left) connected to Odroid XU4 via a FTDI serial to USB chip. Middleground is an OSEPP electronics Robotic Functional Kit with motor controller, distance sensor, etc. Background is OSEPP Tank Kit with motors, treads and chassis hardware. Total cost is around $280.

CONNECTING THE I/O CPU TO THE AI CPU:

The I/O CPU and the AI CPU pass data back and forth via serial communication. To see this part in detail — materials, wiring, and code theory to make a communication link between the two CPUs — visit makezine. com/go/airobotbuild.

3. CREATE THE AI

Now that you are done building the hardware, it's time to create your own AI:

- **Educate yourself:** Read some of the textbook *Artificial Intelligence: A Modern Approach* (Pearson) by Stuart Russell and Peter Norvig. Its 1,000+ pages. It's so thick my friends use it as a monitor stand. Then read a few of the 1,000s of AI white papers published last year. Then forget everything and come up with your own ideas. You most likely noticed there are no smart robots walking around delivering packages or fixing your car. That's because AI needs new and innovative ways to solve real world problems.
- **Invent an AI.** You can start by thinking up a problem. Then dream up a solution and implement it. Keep trying until you get it to work. Then think up a more difficult problem. Having your robot get you a Vitaminwater or other beverage from the fridge is a classic problem. Devise a strategy. Write a program. Test it. BTW, if you solve the "get a beverage" problem, you may become fairly wealthy.

Start simple — the key is to get an understanding of how the AI works with the hardware. Use one sensor and have the robot react to data from it. From there, keep adding sensors of various types. Use multiple sensors simultaneously. A sensor can be as simple as a switch on a bumper to detect hitting walls. Lidar range finding is very promising, try it if you can afford the price. Experience is everything when working with sensors.

FROM THERE, YOU CAN:

- Use OpenCV with a webcam, try all the cool features, like edge detection and face detection.
- Read up on neural networks, they're very popular but time-consuming to implement. They learn tasks by considering large numbers of examples.
- Make a "hand" and put it on you robot. There are few commercially available grippers that can pick up a variety of objects. Try a new approach here. If it works, contact Amazon and see if their warehouse robots "need a hand" (sorry, bad joke).
- Try new approaches with the AI. I worked at a startup that believed in biomimicry (we modeled the AI and mechanical robot systems on the human body). A friend just designed a chip that has an analog neural network.
- Form a club. Have an AI contest. Get your friends to compete. Make it fun like DIYrobocars: diyrobocars.com.

And most of all, have fun and keep experimenting. ⊘

SOME OF MY AI ROBOTS USING MY AI ROBOT BUILDING TECHNIQUES:

AIV3 ROBOT
(Figure **I**)

Purpose: Compete in AI robot combat contests
AI code: Neural networks using Google's TensorFlow
Kit based on: OSEPP Triangular Tank, much hacked
I/O CPU: mbed LPC 1768
AI CPU: Odroid XU4, 8 cores, USB3, HDMI
Cameras: 2x Logitech C920 webcams with wide-angle lenses
Weapon Arm servos: 2x Dynamixel from Robotis
Interface to Dynamixels: USB2AX
OpenCV: Yes
I/O code: C++ using the mbed IDE

AIV3 electronics (Figure **J**) with Odroid XU4 (with white fan), mbed LPC 1768 CPU (blue), and FTDI circuit board (red)

AIV4 ROBOT
(Figure **K**)

Purpose: Compete in AI robot combat contests
AI code: Strategy and tactics reinforcement learning
Kit based on: OSEPP Triangular Tank, much hacked
I/O CPU: mbed LPC 1768
AI CPU: Odroid XU4, 8 cores, USB3, HDMI
Cameras: 2x Logitech C920 webcams with wide-angle lenses
Weapon Arm servos: 2x Dynamixel from Robotis
Interface to Dynamixels: USB2AX
OpenCV: Yes
I/O code: C++ using the mbed IDE

AIV4 (Figure **L**) ready for "play"

Metal *meets* Mettle

LISA WINTER USES DECADES OF BOT BATTLE EXPERIENCE TO HELP SAVE THE WORLD

Written by DC Denison

Lisa Winter has been a roboticist since she was 10 years old. Now 31, she has competed in all U.S. Robot Wars and BattleBots competitions since 1996, including ABC's BattleBots Season 1 and 2. In 2012, Lisa cofounded a smart toy company, Robot 11, building smart wearables and Bluetooth connected toys. As Engineering Project Manager at Mattel, she helped launch Fisher Price's first and only wearable baby monitor. Lisa currently leads a R&D hardware team at Huawei.

How did fighting robots prepare you for jobs in hardware?
It gave me years of experience managing projects. At age 10, I was personally responsible for a Robot Wars robot. Later that included not only design and construction, but also deadlines, finances, and social media. That was a lot of pressure. All that experience was perfect for becoming a project manager. Not the normal route, but it was legitimate experience and from a very young age.

Has your experience with battlin' robots influenced the way you think that robots will integrate into our daily lives?
Robots are everywhere now, and we don't really notice them much, which is cool. Most people have a handful or more in their home already, depending on your definition of 'robot.' There are robotic vacuum cleaners, self-driving cars, washing machines, etc. It's hard to predict where robots will show up next, but that's because they are going to be everywhere, and it probably won't be too noticeable or scary — just common life. No one feels awkward having a dishwasher that has robotic features. They're not scared that the robotic dishwasher in their kitchen is going to attack them in the middle of the night. That robot has been assimilated. We're fine with it.

Do people ever ask you what fighting robots and cute baby monitors have in common?
Often people asked if I was building killer robots for Mattel. I said no, I'm trying to save lives! The baby monitor is a product that could literally do just that.

Robot Wars also has a positive side — one of the founding principles of Robot Wars is to channel fighting into robot-versus-robot, instead of humans getting hurt.

You and your dad are doing a competition with robots and AI?
Yes, ever since I've been building robots with my dad, we've had this sort of friendly argument. He says, "If I put an AI computer brain in my robot, it would be so much faster, it would win every competition." His idea is that a robot with AI could spot the other robot and turn really quick, much faster than a human can think or react, and it won't turn too much. Humans tend to overcorrect, for example. However, I'm still going to continue driving my robot because I love doing that. So now it's come down to: "OK, see you on the court. Build this AI robot and whenever you do, I'll fight it."

And that's happening now?
Yes, we're actually doing that. I built a few robots that I can control with a transmitter, and he built a few that are AI controlled. We're calling it "AI or Die." The goal is to have a contest in the Bay Area. All robots with functional AI are encouraged to contact us at AiOrDie.tv

You juggle so many projects.
It's just my personality. Even lately I'm saying, "I've got to find more stuff to do." I need that pressure. If the project interests me then I'll make time for it.

Something I'm incredibly excited about is my collaboration with the Sumatran Orangutan Conservation Programme (SOCP). We have a small team, and together our goal is to track the orangutan's health and vitality, map their home territory, and use this information to maximize their health outcomes, and create a protected habitat free from deforestation and illegal trade.

The possibility of helping an endangered species keeps me extremely motivated.

Hep Svadja, Mike Winter

Any career advice for people in the robot combat world?
The really cool thing is that robotics can branch into almost any type of job. I see that now as I'm mentoring a robotics team at Berkeley High School. Teaching a class for kids is a new experience for me, but it's been really eye-opening. Some kids are interested in learning fundraising — asking companies to donate money, or applying for grants. Others are really interested in the building aspect. Some really want to learn how to use the tools that we have. It's awesome because I can see myself in them: you have to be the fundraiser, you have to be the builder, you have to be the 3D CAD designer, you have to do the social media. If you can do all that, and handle the stress of doing it in a very short time, then it's really good experience for whatever job you want to get into. 🌀

DC DENISON is the co-editor of the Maker Pro Newsletter, which covers the intersection of makers and business, and is the senior editor, technology at Acquia.

You can follow Lisa on her **YouTube channel**: youtube.com/lisawinterx
and her **Web page**: lisaxwinter.com
And keep an eye out for the upcoming **"AI or Die" competition**: AiOrDie.tv

Written and photographed by Greg Voronin

Rochambeau Bot

TIME REQUIRED:
A Weekend

DIFFICULTY:
Intermediate

COST:
$250

MATERIALS
» **Mycroft Mark 1 voice assistant** mycroft.ai
» **Raspberry Pi Camera Module v2** Adafruit #3099, adafruit.com
» **Ribbon cable for Pi Camera, 24" or longer** such as Adafruit #1731

TOOLS
» **Torx T10 screwdriver, bit, or wrench**
» **Computer** Find the Clarifai API at clarifai.com/developer, and get the project code at github.com/lachendeKatze/skill-rock-paper-scissors.
» **3D printer** to print Pi Camera enclosure. I used thingiverse.com/thing:2379487.

GREG VORONIN is inspired by his kids to learn about all sorts of maker stuff and share what he learns with them and the maker community.

THROW DOWN WITH YOUR MYCROFT VOICE ASSISTANT USING CLARIFAI'S IMAGE AI SERVICE

WHEN THE HOSTS OF THE PODCAST *STARTUP HUSTLE* suggested to Mycroft AI's CEO Joshua Montgomery that his company's Mycroft Mark 1 voice assistant should play a game of rock-paper-scissors, I knew I had to try with mine.

The Mycroft Mark 1 is open source hardware and software that emphasizes the privacy and security of user data. Centered around the Raspberry Pi 3, it is supported by a friendly and responsive community.

By attaching a Pi Camera Module to the Mark 1 and enabling a Mycroft voice skill with Clarifai, an AI company that provides computer vision as an online service through a Python API, I was able to play a game of rock-paper-scissors (RPS) with Mycroft in just one weekend. This is how I did it.

ADDING THE PI CAMERA

After carefully removing the bottom of the Mark 1 enclosure with a Torx screwdriver (Figure Ⓐ), I detached the Raspberry Pi from its posts, flipped it over, and used a 2-foot ribbon cable to secure it to the camera connector (Figure Ⓑ). I reattached the Pi, fed the cable through one of the slots in the enclosure bottom, then closed it back up (Figure Ⓒ). I gently seated the Pi Camera to the other end of the cable. For convenience, I mounted the camera in a 3D printed stand (Figure Ⓓ) that I found on Thingiverse: PiCamera 2 Axis Rotating Holder by migrassi (thingiverse.com/thing:2379487).

The Mark 1 houses a "headless" version of the Pi, which means it doesn't have a screen or keyboard attached to interact

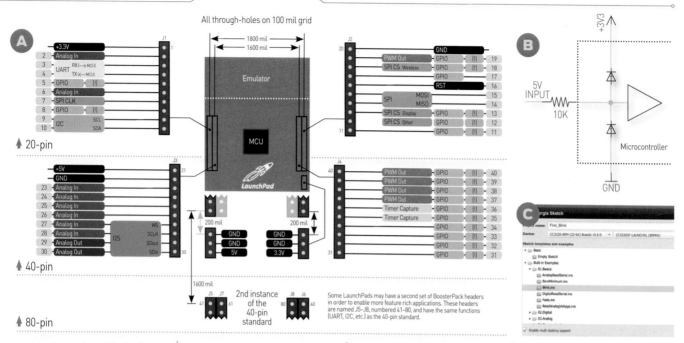

A 20-pin

A 40-pin

A 80-pin

All through-holes on 100 mil grid

Some LaunchPads may have a second set of BoosterPack headers in order to enable more feature rich applications. These headers are named J5–J8, numbered 41–80, and have the same functions (UART, I2C, etc.) as the 40-pin standard.

resistor in series with the 5V signal before it reaches the 3.3V IC (Figure **B**). This limits the current and allows the diode to clamp the voltage to ~4V which the device can handle.

Driving 3.3V → 5V: Because the threshold values used to detect 0/1 voltages are low enough, most 5V systems will register 3.3V as a 1, so no changes are needed to talk to a 5V system using 3.3V.

SOFTWARE

Energia: Energia is an open source IDE for the LaunchPad boards. LaunchPads may not be Arduinos, but they can run the same sketches as an Arduino when you use Energia.

How's that possible? Every microcontroller is different, so embedded software developers often create a Hardware Abstraction Layer (HAL) as the foundation of any project. This allows the software application they write to be portable (to be used on a different microcontroller). The Arduino HAL is all the functions you can call, like `digitalWrite` and `analogRead`. Energia recreates these functions on the

LaunchPad, so Arduino projects can be compiled and run on the LaunchPad boards.

> **NOTE:** Not all LaunchPads are supported in Energia. Check their website before picking a board to use for a project.

CCS and CCS Cloud: Code Composer Studio is TI's full professional development environment for their microcontrollers. It's a very powerful tool but has a steep learning curve, so I wouldn't recommend it for the average maker. Instead, check out CCS Cloud — a browser-based IDE for TI devices and LaunchPads. It has all the functionality of Energia and allows you to debug your code.

CHOOSING A LAUNCHPAD FOR YOUR PROJECT

There are 32 different LaunchPads listed on TI's website. It's nice to have options, but choosing one for your project can be confusing. Let's blast away that befuddlement!

There are four families of LaunchPad: SimpleLink (wireless communication), MSP430 (ultra low power),

C2000 (real-time controls), and Hercules (safety). Support for the C2000 and Hercules boards in the maker community isn't as well-developed, so we'll focus on the MSP430 and SimpleLink boards.

SIMPLELINK

For Wi-Fi, Bluetooth, and other wireless applications, check out the SimpleLink family. The latest and greatest is the CC3220SF LaunchPad, which supports 802.11 b/g/n networking. At just over $50, this is one of the more expensive LaunchPads, but is a great fit for nearly any maker application.

MSP430

MSP430 devices are what started the LaunchPad family. They're the perfect mix of processing power and cost. My favorite is the classic MSP430F5529 LaunchPad, which is only $13. More recently the MSP430 family has expanded into the 32-bit world with the addition of the MSP432 devices. These utilize an ARM Cortex M4F processor and are a great fit for applications needing a little more processing power. And they too cost only $13.

GETTING STARTED

Now that you have your LaunchPad, it's flashing time! Keep your clothes on — in this article we'll be flashing LEDs. Go to dev.ti.com and select CCS Cloud. You'll need a TI account to access the tool, but it's easy to register and totally free.

Once you're in, you'll be greeted with a blank IDE. Click File → New Energia Sketch. Name your sketch (I chose *First_Blink*), and select the board you're using. Expand the folder *Built-in Examples → 01.Basics*, and select the *Blink. ino* sketch. Then click on Finish (Figure **C**). You've created your first project.

Now run it on the LaunchPad. Plug your LaunchPad into your computer using a USB cable. Allow any drivers to install, then click the green Run button in CCS Cloud. The IDE will compile your code and download it to your LaunchPad. In a few seconds, you'll see an LED blinking on the board. Easy!

Now that you're up and running, there's much more to explore. Try the other examples, modify them, debug your work, and hook up external hardware to make your own projects. *◎*

Brains On Board

AIY Voice Kit

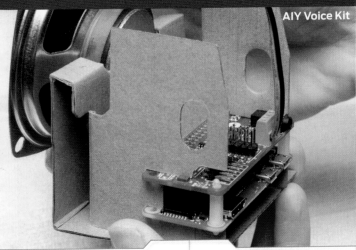

Remind me to buy Kristine a birthday present on May 8th

Alright, I will remind you on May 8th 2018 at 8am

THESE THREE KITS CAN HELP CONNECT YOUR ELECTRONICS PROJECT TO POWERFUL AI

Written by Kelly Egan

KELLY EGAN is an artist, teacher and creative coder living in Providence, Rhode Island. He is a founding member of both Baltimore Node and Ocean State Maker Mill.

MAYBE YOU HAVE BUILT A REMOTE PET FEEDER, or an automatic lock for your front door, or a robot — all good maker projects. But there is usually one thing missing that would make them great: a little bit of intelligence. What if your pet feeder knew to feed the cat and not the squirrels, or you could tell your door to unlock when your hands are full? And who hasn't wanted a robot you could talk to? AI can make a good project amazing. Here are some new kits that let you capitalize on the power of AI in your projects.

AIY VOICE

aiyprojects.withgoogle.com/voice
The heart of the Google AIY Voice setup is the Voice Hat, an add-on board for the Raspberry Pi Zero. The Voice Hat doesn't actually have any onboard speech processing. That is handled by Google's cloud (or another service like Amazon's Alexa). The Hat primarily provides a decent speaker amplifier and a secondary board with a stereo microphone.

The original Voice Hat, using a full-size Pi, broke out a number of input-output pins with space to easily add servos as well as drive some higher loads (up to 500mA). This makes it easy to give your voice-controlled device some motion or other physical interface. The newer release uses the Pi Zero, so it's a little more limited — but it does come with the Pi Zero rather than having to supply your own.

AIY VISION

aiyprojects.withgoogle.com/vision
Like the Voice kit, the AIY Vision kit also has an add-on board for the Raspberry Pi Zero, a cardboard enclosure, and an arcade button. But this board, the VisionBonnet, has some real power to do onboard image analysis without the cloud. It uses an Intel Movidius MA2450 vision chip along with the Raspberry Pi Camera Module.

The MA2450 is designed for low-power environments like mobile phones and helps the Pi deal with the vast amount of data generated by the camera's live video stream, allowing this small device to process the input and recognize faces and other objects quickly.

Google's example code provides pre-trained models for faces, expressions, and objects like cats and dogs. You can even train your own models, though not on the device itself. For that, you'll need to dive into a deep learning environment

AIY Vision Kit

79.82%
Fuji
Category: Edible fruit

20.80%
Orange
Category: Edible fruit

Matrix Voice

like Google's TensorFlow. The process of classifying what an object is, from thousands of images, is too intense for such a small device to do in a reasonable amount of time. However, the raw image processing of the device is still powerful and really useful if you want to make a responsive vision-based interface without an expensive computer and graphics card attached.

The form factor of the Pi Zero doesn't allow for the additional breakouts found on the Voice Hat, like the transistors to drive high loads, but it does break out four of the Pi's I/O pins, power, and ground so you can connect additional inputs and outputs. You may also eventually want to create a sturdier enclosure, as the included cardboard setup can wear out after a few reassemblies.

MATRIX VOICE
matrix.one/products/voice

The Matrix Voice is the most capable of the three boards with an 8-channel microphone array and a chip for audio processing. This is the second board by Matrix Labs, preceded by the more expensive but more full-featured Matrix Creator.

The Matrix boards use a Field Programmable Gate Array (FPGA) to process the raw audio input from the 8-channel microphone array by performing tasks such as noise cancellation and beamforming. Matrix has preprogrammed the FPGA with many of the needed audio algorithms but you are free to tinker with them. Like the AIY Voice kit, the speech recognition and natural language processing used to turn users' speech into useable commands is

handled by cloud services like Google or Amazon.

The Matrix Voice supports a few more features than the AIY Voice, with both speaker output and headphone jack, an LED ring, and additional I/O pins. If you get the version with the ESP32 chip, you can operate the board with or without a Raspberry Pi.

Matrix Labs sees their boards as part of a platform of IoT devices and apps, and they've even provided a repository so you can easily add other people's apps to your Matrix-enabled Pi.

• • •

Using a voice assistant such as Google Assistant or Amazon Alexa with either the AIY Voice or the Matrix Voice requires some significant setup with those services. You'll need to answer questions about the

app you're creating, as well as create tokens and credentials that connect your device, app, and the different cloud services. This process is documented but not particularly straightforward.

In addition, there's some setup required on the Pi itself to configure the hardware and install the development environments and examples. It's useful to have some experience with Linux and/or the Raspberry Pi environment if the build doesn't go completely smoothly.

The big advantage of all these boards is the preprocessing they do with raw audio and video input. And with the audio boards, many of the AI features of the cloud services like Google Assistant and Amazon Alexa can be accessed from a simple Raspberry Pi computer. So why not give your next project some smarts? ●

AN ARRAY OF SENSORS AND A POWERFUL
BUT MAKER-FRIENDLY BRAIN LET THE
SKYDIO R1 TRACK YOU ANYWHERE

Written by Mike Senese

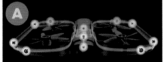

Follow the Leader

MULTIROTOR DRONE MAKERS HAVE LONG PROMISED EVERYONE THEIR OWN PERSONAL FLYING CAMERA, something that could autonomously follow overhead, determining the best angles and maneuvers to get professional-caliber footage. Part of that has come true: Drone computer vision software has enabled follow-me functions, and users are now able to set various shooting styles on their aircraft too. But full spatial awareness — the ability to identify and avoid obstacles in any direction — has thus far only received limited activation, meaning chances of an autonomous-flight crash are still very realistic.

Enter the Skydio R1, a new quadcopter that employs powerful AI processing and a bevy of visual sensors to track and film its subject while flying itself through almost any environment — through a forest,

underneath overhangs, against brick walls, and more. *Make:* got to test it out and it works surprisingly well, following as we sprinted through a crowded yard behind the Skydio offices in Redwood City, California, gracefully slaloming itself around trees, posts, and an elevated scissor lift.

"THE MAGIC OF THIS PRODUCT IS IN THE SOFTWARE."
— SKYDIO CEO ADAM BRY

The R1 rapidly crunches the visual data of 12 onboard cameras, set up as six stereo pairs (Figure Ⓐ), to create a three-dimensional map of everything in its surroundings, while tracking its subject with a 13th camera, which also records in 4K video. As it follows along, it constantly adjusts its

flight path to go around, over, or under any obstructions, even if flying sideways or backward.

For the necessary processing power, each R1 employs a powerful Nvidia Jetson TX1, a board that is focused on visual AI applications (and comes in a dev kit configuration for makers). But Skydio feels the real work comes from what goes inside. "The magic of this product is really in the software," Skydio co-founder/CEO Adam Bry says.

Bry and his two partners met as grad students at MIT before transitioning to Google to help create the Project Wing division, focused on experimental flight initiatives. They departed to begin Skydio in 2014, prototyping and refining their hardware until the R1's debut in early 2018.

Costing $2,500, the R1 isn't a toy, but rather, a powerful computer that just

happens to fly. And while the first product is focused on selfies for the solitary sportsperson, the technology inside has the potential for industrial applications and so much more. ◕

"Voxel view" gives a peek into the R1's method of mapping obstacles.

[+] Go inside the Skydio HQ with *Make:* at makezine.com/go/skydio

MIKE SENESE is the executive editor of *Make:*

MAKE sure it's in your toolbox.

SOPHY WONG
is a designer and
maker whose projects
range from period
costumes to Arduino-
driven wearable tech.
She can be found
at sophywong.com
and on her YouTube
channel chronicling
her adventures in
making.

Written and photographed by Sophy Wong

Photo Op **SelfieBot**

Shoot and instantly print images with this adorable camera

Instead of a selfie booth, bring SelfieBot to your next party! This Raspberry Pi project has a mind of its own: SelfieBot giggles, snoozes, and prints selfies on its thermal printer at the push of a button. Animated facial expressions and sounds are triggered by movement-based interactions, thanks to a tiny accelerometer board inside.

My husband and I built our original SelfieBot for our booth at Seattle Mini Maker Faire 2017. This battery-powered version is a simpler build, contained in a customizable laser-cut acrylic case. We 3D printed the walls of the case, but you can also laser-cut them or build them any way you like. Use the animations and sounds provided, or create your own!

This is a big build, and definitely a multiday project with a lot of parts and several fabrication techniques. It's a fun mix of electronics and fabrication, and there are lots of opportunities to customize it along the way.

At the heart of the circuit is the Pi Cobbler add-on board, which routes all the pins of the Raspberry Pi neatly to a PermaProto circuit board for connecting the accelerometer, printer, speaker and amplifier, and various other bits (Figure Ⓐ).

Here's an overview of the project; for the complete instructions visit the project page online at learn.adafruit.com/raspberry-pi-selfie-bot.

1. LASER CUTTING

You'll start by building the case. It's made up of three laser-cut acrylic panels separated by two 3D-printed wall sections, all held together with screws and rivet nuts.

Download the panel files from the project page and cut them in ⅛" acrylic on your laser cutter. Leave the protective film on your acrylic while cutting for clean panels with no fogging. For the top and middle panels, there are also optional engraving files, so your holes will be nicely labeled and easy to navigate while building (Figure Ⓑ).

2. 3D PRINTING

Use the provided files to 3D print the front and back wall sections (Figure Ⓒ) and the two handles. You'll probably need to cut each wall into two pieces to fit on your printer bed. (You can also laser-cut the walls from acrylic or even cardboard; we've provided files for that too.)

TIME REQUIRED:
A Few Weekends

DIFFICULTY:
Intermediate

COST:
$225–$275

MATERIALS
MAIN COMPONENTS:
» **Raspberry Pi 2 single-board computer with microSD card, minimum 16GB**
» **Raspberry Pi Camera board v2**
» **HDMI 5" Display Backpack** Adafruit #2232, adafruit.com
» **Pi Cobbler breakout board and cable** Adafruit #2029
» **ADXL345 accelerometer board** Adafruit #1231
» **Mini thermal receipt printer** Adafruit #597
» **Mini metal speaker, 8Ω** Adafruit #1890
» **PAM8302 audio amplifier board** Adafruit #2130
» **Arcade button with LED** Adafruit #3429
» **UBEC step-down converter, 5V 3A output** Adafruit #1385
» **PermaProto Breadboard PCB, half size** Adafruit #1609
» **Battery, NiMH (for R/C cars) 1,600mAh 7.2V, with compatible connectors**
» **Wireless mouse and keyboard** Buy a cheap set and leave the USB dongles plugged into your Pi permanently.

CONNECTORS ETC.:
» **USB connector shell** Adafruit #1389
» **Right-angle HDMI adapter**
» **12" HDMI cable with small plugs** Adafruit #2420
» **Micro-USB cables (2)** at least one with a left-angle micro-USB connector

» **4-pin JST cable connector set** Adafruit #578
» **Power switch, 2A** such as AllElectronics #RS-223, allelectronics.com
» **Stranded wire, 22 or 24 AWG** in assorted colors
» **Silicone stranded wire, 16 AWG** in black and red
» **Heat-shrink tubing** various sizes and colors

FOR CASE AND ASSEMBLY:
» **Acrylic sheets, 3mm (⅛") thick, at least 9½"×12½" (3)**
» **3D printer filament**
» **Nylon spacers: ½" (4) and ¾" (4)**
» **Grip rivet nuts, #6-32 (6) drill size 12,** McMaster #93482A605, mcmaster.com
» **Machine screws, #6-32, 2½" long (6) and matching washers** for case
» **Machine screws: #4-40, ¾" long (6) and matching nuts** for handles
» **Filler and spray paint (optional)**
» **Model Magic clay or Blue Tac (optional)** for masking holes
» **Craft foam, 3mm**
» **Gaffer tape**
» **Velcro tape, self-adhesive, 1" wide** industrial strength
» **Double-stick foam tape**

TOOLS
» **Laser cutter and 3D printer (optional)** Check with your local makerspace for access to these tools, or send the files out to a service.
» **Wire cutters / strippers**
» **Soldering iron and solder**
» **Screwdriver**
» **Handsaw**
» **Drill**
» **File**
» **Hot glue gun**
» **Epoxy**
» **Sandpaper**

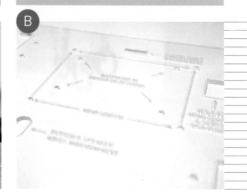

NOTE: Depending on the material you're printing with, allow for shrinkage after cooling by scaling the model up before printing. For ABS, somewhere between 1.5% and 2.5% should work, but I strongly recommend you test and measure to calculate your actual shrinkage factor before printing the whole run of parts.

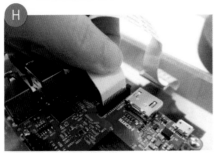

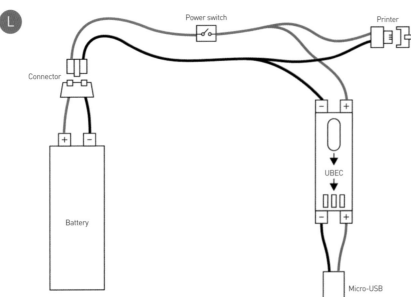

The handles are held together with screws and captured nuts. Heat each nut with your soldering iron and press it gently into its hexagonal hole (Figure D).

3. PAINTING

I chose to leave the back and middle acrylic panels clear, and back-paint the front panel for a clean finish. Leave the protective film on the front of the panel, and paint the backside (Figure E). Spray paint needs to be applied to acrylic in many very light coats, so take your time. I masked and painted the backside in three different colors. Choose your favorite colors and get creative here.

Glue each wall section back together (epoxy works well on ABS), then finish as desired. I sanded, filled, and primed the outside of the walls before painting with spray paint. I left the handles unpainted.

4. PROGRAM THE RASPBERRY PI

Download and unzip the disk image from the project page online, burn it on your microSD card (Adafruit has a good tutorial), then insert the microSD card into the Raspberry Pi, and you should be ready to go.

Thanks to Kim Pimmel for writing this code. The core library being used is PyGame, for running SelfieBot's expressive animations and driving the selfie functionality. The CUPS library is used for printing to the thermal printer. The accelerometer library is adxl345.py written by Jonathan Williamson.

The animated facial expressions (Figure F) and sounds are what make this project special. Here's how they were made, in case you want to customize your SelfieBot's personality.

Faces — are a series of images that play in a loop when triggered. Keeping the basic face simple made it easy to create expressions by changing the shape of the mouth and focus of the eyes. Each frame was drawn in Adobe Illustrator and saved as a *.png* file.

Sounds — are my voice, recorded as individual *.wav* files and processed in Reason for a robo-cyborg effect.

5. BEGIN ASSEMBLY

Mount the thermal printer, arcade button, and Pi Camera in their holes in the front panel (Figure G). Stack the front wall section on top.

On the middle panel, mount the Raspberry Pi to the backside, and the HDMI Backpack screen and the speaker to the front (I made a speaker housing from craft foam). Add the panel to the stack, slide the cables through their holes, and connect the camera to the Pi (Figure **H**).

6. SET UP THE COBBLER CIRCUIT

Insert the Cobbler into the center of the PermaProto board and solder all pins in place (Figure **I**). Connect 5V to one rail of the board and 3V to the other (Figure **J**). (To remind myself of the different voltage rails, I marked the 3V wire with a piece of white heat-shrink.) Connect ground to both ground rails.

Now connect the main components as follows:

Thermal Printer	RX to GPIO 14, and related ground to GND rail
Arcade Button	Data pin to GPIO 17, power to 5V rail, ground to GND rail
Accelerometer	SCL to GPIO 3, SDA to GPIO 2, VIN to 3V rail
Amplifier	A- and A+ to audio out on Pi, Power to 5V rail, GND to ground rail (Figure **K**)

7. BUILD THE POWER CIRCUIT

Power and ground from the battery are split: one pair of wires goes directly to the thermal printer, the other pair goes to the input side of the UBEC step-down converter. Power and ground from the output side of the UBEC goes to power the Raspberry Pi via the micro-USB port (Figure **L**). You'll also connect power and ground from the Raspberry Pi GPIO to the PermaProto board.

Mount the power switch in the back wall section, using a drill and file to make a hole to fit your particular switch.

8. PUT IT ALL TOGETHER

Use double-stick foam tape to mount the PermaProto, amp, and accelerometer to the middle panel. Then plug everything in (Figure **M**). The display plugs into the HDMI port on the Raspberry Pi, and is powered by USB from the Pi. The speaker wires connect to the screw terminals on the amplifier. See the project page online for all the connections.

To keep the battery in place (yet

removable), affix it to the bottom of the panel with industrial-strength sticky velcro.

9. FINISHING TOUCHES

Power up your SelfieBot and test it. When you're sure everything is working, reward yourself by pulling the protective plastic sheet off the screen (so satisfying!), and then close up the case with the 2½" screws and rivet nuts.

Finally, mount the handles on the case, inserting the ¾" screws from the back piece and screwing them into the captured nuts in the front piece (Figure **N**).

Congratulations, you're finished building SelfieBot!

FUN WITH SELFIEBOT

Load up the printer with paper, flip that switch, and get to know your new friend!

TO TURN ON:

Flipping the power switch on will boot the Raspberry Pi (Figure **O**). You'll see the Raspberry Pi boot screen, the arcade button will illuminate, and the green light on the printer will flash intermittently. Open the SelfieBot program on the desktop and run the program in IDLE.

SelfieBot reacts to motion. When shaken gently or tipped, SelfieBot giggles, as if tickled (Figure **P**). When placed on its back, SelfieBot "goes to sleep" — closes its eyes and begins to snore (Figure **Q**). When picked up again, SelfieBot wakes up, smiles, and emits a cute vocal bloop.

TO TAKE A SELFIE:

Press the arcade button to enter selfie mode. Smile at the camera, and press the button again to take and print a photo.

TO TURN OFF:

To exit the SelfieBot program, press Esc on the keyboard. Shut down the Raspberry Pi and then flip the power switch off. ◓

Visit the project page online at learn. adafruit.com/raspberry-pi-selfie-bot for complete assembly instructions, alternate build methods, project code, and files for cutting and printing.

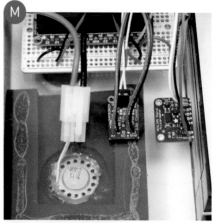

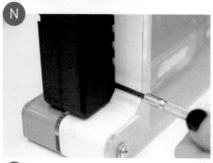

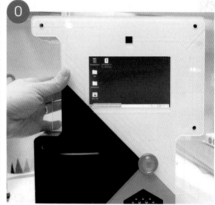

TIME REQUIRED:
3–4 Hours

DIFFICULTY:
Easy

COST:
$30–$50

MATERIALS
» **Raspberry Pi Zero W single-board computer**
» **Monochrome OLED display, 128×64** Adafruit #326, adafruit.com
» **Hookup wire, 28ga or smaller**
» **Socket screws, M2.5×10mm (4)**
» **MicroSD card, 8GB**
» **USB power supply, 1A** for in-car use
» **Enclosure** Make your own or print my files

TOOLS
» **Computer**
» **Soldering iron**
» **Wire cutters/strippers**
» **Hex wrench**
» **3D printer (optional)** Visit makezine.com/go/digifab-access to find a machine or service.

Skim Scam Scanner

Written by Tyler Winegarner

Build a dashboard gadget that scans constantly for potential gas pump card skimmers

What's worse than gazing over your credit card statement and seeing payments you didn't authorize? You frantically mark up your statement with a red pen before you make that call to your bank, hoping that they'll reinstate the funds after freezing your card!

The root of this vulnerability is the completely insecure nature of credit cards themselves. The traditional magnetic stripe on your card contains all the data necessary to complete a transaction, and because it's not encrypted, this data can easily be stolen using simple electronics. So-called "card skimmer" devices deployed by crooks act like a "man-in-the-middle," intercepting and recording your credit card data before passing it along to the point-of-sale machine, like a gas station fuel pump.

Last year, Nathan Seidle of SparkFun Electronics did a technical deep-dive

of credit card skimmers that had been extracted from fuel pumps by his local police force. The result was an app, released for Android and iPhone, that will scan for Bluetooth devices matching the fingerprint of the skimmers he researched.

That's all well and good, but I'm lazy, and slow to develop beneficial habits. I wanted a dedicated, ambient appliance that would always be on the lookout for suspicious Bluetooth devices at the gas station, and would shout at me before I did something stupid with my financial future. However, credit where credit is due, this project wouldn't exist without Seidle's hard work.

1. HARDWARE
There are only two components for this build, a Raspberry Pi Zero W and the Adafruit SSD1306 OLED display. First you'll solder the connections for the OLED. When

we package it up, you don't want a lot of slack in the wires, so keep them as short as possible, and use something with plenty of flex: solid 28-gauge wire keeps things manageable, and stranded wire makes it even easier. The SSD1306 supports two different communication protocols: I^2C and SPI. We'll be using the latter — it uses a few more wires, but it's faster. Wire everything up according to this diagram (Figure A).

2. SET UP RASPBIAN
Getting a monitor and keyboard connected to a Pi Zero requires a ton of extra hardware, so I always like to start off with a nice headless setup. The easiest way to do this right from the start is to use the delightful block-based Raspberry Pi setup tool, Pi Bakery. It lets you create a bootable Raspbian image for your Pi that's already set up for your home network and will allow

connections to it via SSH. Here's what my typical setup looks like (Figure **B**).

3. PYTHON AND LIBRARIES

Once you have Raspbian loaded, boot it up and connect to it via SSH. Go into raspi-config, set up something other than the default password, and enable SPI under Interfacing Options. Then install Python with the following command:

```
sudo apt-get install python
python-pip
```

Install the necessary Bluetooth tools with this command:

```
sudo apt-get install bluetooth
libbluetooth-dev
```

Our program runs on Python, so you'll need to get the Bluetooth toolset for Python:

```
sudo pip install pybluez
```

Next, you need to get the software to drive the Adafruit OLED panel:

```
sudo pip install RPi.GPIO
sudo pip install Adafruit_BBIO
sudo apt-get install python-
imaging python-smbus
```

And some specific libraries from Adafruit for the OLED display:

```
sudo apt-get install git
git clone https://github.com/
adafruit/Adafruit_Python_SSD1306.
git
    cd Adafruit_Python_SSD1306
    sudo python setup.py install
```

4. SKIMMER SCANNER CODE

Finally, download our scanning software:

```
cd ~
git clone https://github.com/
photoresistor/raspi_skimscan
```

And give it a test run by doing the following:

```
cd raspi_skimscan
python raspi_skimscan.py
```

You should see "Scanning…" with a scrolling ellipsis that indicates a fresh scan every 10 seconds. If you have the ability to change the Bluetooth name of your mobile device, you can test your scanner by changing your mobile's name to "HC-05" and verifying that you see a warning message (Figure **C**). When you're done, quit by pressing Ctrl-C.

The last thing to do is set up the Pi so it

runs our script at startup. Make your Python script executable with this command:

```
sudo chmod +x raspi_skimscan.py
```

Then, set up **rc.local** to launch the script on boot:

```
sudo nano /etc/rc.local
```

And add the following line to that file, directly above the line that says **exit 0**:

```
sudo python /home/pi/raspi_
skimscan/raspi_skimscan.py &
```

That's it! You're done. Reboot your Pi and verify that everything is running right.

5. ASSEMBLY

If you haven't already, 3D-print the two files for the enclosure. You can download them from my GitHub repo at github.com/photoresistor/raspi_skimscan.

Place the OLED screen over the mounting posts in the enclosure's top piece. If you like, you can secure it with tiny dabs of hot glue. Place the Pi over the offsets in the bottom piece (Figure **D**), and close it up. Secure the two halves together with M2.5×10mm cap head screws.

Attach the enclosure to your car's dashboard or console, anywhere visible but not distracting. Power it from any USB power source in your car, as long as it provides at least 1 amp.

SCANNING FOR SKIMMERS

Now wait a minute! If you compare my code to what Nathan Seidle wrote for his mobile app, you'll realize that mine is a bit less stringent. It reports any Bluetooth devices that identify by the same name as those commonly used in gas pump skimmers — namely HC-03, HC-05, or HC-06 — but doesn't ping them for a reply. (I didn't do the secondary verification like Seidle does because I wasn't able to get hold of any actual skimmers to test against.) These are cheap Bluetooth dev boards that are also used in legitimate products and DIY projects — so it's entirely possible that this could introduce a few false positives.

If your scanner reports a hit, look around. There's a good chance a criminal put a skimmer in that pump. Or maybe there's a kindred maker hanging around. Hacker types are usually easy to spot — just make sure to check the color of their hats before you fuel up. ◉

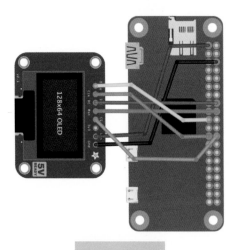

OLED	RaspPi
Data	MOSI
Clk	SCLK
DC	GPIO 23
Rst	GPIO 24
CS	CE0
Vin	3.3V
Gnd	Ground

A

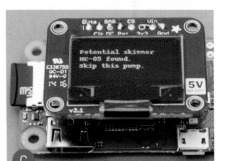

B

C

D

TYLER WINEGARNER, video producer for *Make:*, is a maker, tool user, story teller, and skill hoarder. He is driven by the weird and wonderful.

Hue've Got the Look

Control the color of these LED party shoes and matching bracelet wirelessly with Adafruit's handy app

Written by Angela Sheehan

Angela demonstrates her vivid color-changing shoes and bracelet with this long exposure shot.

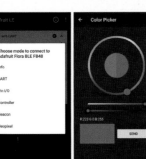

ANGELA SHEEHAN is a maker and educator passionate about DIY electronics, costuming, and craft tech. She has been tinkering with wearables since 2005. Find her projects on GellaCraft.com, Twitter @the_gella, and Instagram @gellacraft.

I'm one of those people who is always late to a party because I can't decide what to wear. I'm also someone who creates elaborate last-minute projects for those parties. This project — I call them my Insta-Hue Heels — was designed and created in less than a week for me to wear to the SparkFun Electronics holiday party.

LEDs on the heels are wirelessly controlled by a matching bracelet connected to Adafruit's free Bluefruit LE Connect app, allowing for quick color changes for any occasion. The magical effect of instantly controlling color through a smartphone is a great way to interact with other partygoers — let them try changing your look!

It's easy to replicate with a little time, patience, and the perfect pair of pumps.

1. CODE AND FUNCTION OVERVIEW

Setting up the Bluetooth connection was easy using Adafruit's tutorials. I started with the framework from their *Controller* example for the Bluefruit LE board. The bracelet uses the Color Picker tool built into Adafruit's Bluefruit LE app to control the LEDs. Since the app (for Android or iOS) will only pair to one Bluetooth device at a time, I used inexpensive, short-range RFM69 transceivers to broadcast the color info wirelessly from the bracelet to both heels. This allows control of multiple costume pieces at a time without writing a ton of custom software. I used SparkFun's well-documented RFM69 hookup guide for code examples to integrate the additional radio broadcasting function into the project.

Upon startup, the Flora Bluefruit module in the bracelet waits to connect to the Bluefruit app. After pairing, opening the

app's Controller menu and the Color Picker function brings up a color wheel for choosing a color on the phone's screen. Hitting the Send button sends the color values (RGB) over Bluetooth to the bracelet (Figure Ⓐ).

The bracelet stores the received values in variables, then retransmits the received color to the heels via the RFM69 radio transceiver. It also displays the color on the bracelet's NeoPixel ring using Adafruit's NeoPixel library.

The heels receive the broadcast values and update the color displayed on the flexible LED panels with the `ColorWipe()` function from the NeoPixel library. Depending on manufacturer of your panels, you may need to adjust the settings in code to set the origin pixel of the matrix (Figure Ⓑ).

More detail on the code and my latest prototype is available for download at github.com/GellaCraft/InstaHueHeels.

> **NOTE:** These LED panels can draw a significant amount of current. The project code sets the brightness of the panels quite low using the NeoPixel library, in order to lower the current draw and extend battery life. For best practices using NeoPixel matrices, refer to Adafruit's NeoPixel Uberguide at learn.adafruit.com/adafruit-neopixel-uberguide.

2. CHOOSING SHOES

Careful selection of the shoes made it easy to integrate and hide the electronics. I found 3" block heels that nicely accommodate the flexible LED matrix panels. This height also allows placement of a 1,200mAh LiPo battery on the inside edge with enough clearance to avoid potential damage from walking (Figure Ⓒ).

TIME REQUIRED:
A Weekend

DIFFICULTY:
Intermediate

COST:
$200–$300

MATERIALS
FOR THE BRACELET:
» **Pro Micro Arduino dev board, 3.3V/8MHz** SparkFun Electronics #12587, sparkfun.com
» **RFM69HCW wireless transceiver module, 915MHz** SparkFun #13909
» **Flora Wearable Bluefruit LE Module** Adafruit #2487, adafruit.com
» **NeoPixel Ring, 16×5050 RGB LED** Adafruit #1463
» **Surface-mount navigation switch** SparkFun #8184
» **Battery, lithium ion polymer (LiPo), 3.7V 400mAh** SparkFun #13851
» **LilyPad Simple Power board** SparkFun #11893, or equivalent LiPo connector with switch. I used an older version of this board I had in my studio (SparkFun #10085), but the latest version includes a charging circuit, which is handy.
» **Silicone stranded wire** I used 30AWG.
» **Solid-core hookup wire, 3.07" length** for radio antenna
» **Ribbon, 2" wide**
» **Magnetic closure**
» **Decorative ring** Bring your NeoPixel ring to the craft store to find something from the jewelry aisle that can accommodate, or fabricate your own.
» **Acrylic sheet, ¼" thick (optional)** if you're fabricating the casing for the components on the bracelet

FOR THE SHOES:
» **Pair of heels with 3" or higher block heel** See Step 2 for notes on choosing shoes.
» **Qduino Mini Arduino dev boards (2)** SparkFun #13614
» **RFM69HCW wireless transceiver modules, 915MHz (2)** SparkFun #13909
» **Flexible 8×8 NeoPixel RGB LED matrices (2)** Adafruit #2612 or Amazon #B07418JSWD
» **Batteries, LiPo, 3.7V 1,200mAh (2)** Adafruit #258
» **Stranded wire, silicone covered** I used 30AWG and 24AWG.
» **Ribbon cable (optional)** to connect the RFM69s to the Qduinos
» **Solid-core hookup wire, 3.07" lengths (2)** for radio antenna
» **E6000 adhesive** or other glue that will adhere to your chosen shoes
» **Gaffer's tape** in a color matching your shoes

TOOLS
» **Bar clamps, about 4" (3 or more)**
» **Soldering iron and solder**
» **Third hand / helping hands tool**
» **Wire cutters / strippers**
» **Heat gun (optional)**
» **Laser cutter**
» **Hot glue gun**

Geoff Decker – Hidden Vision Photography, Angela Sheehan

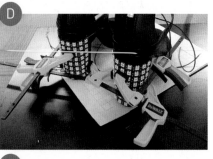

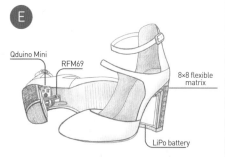

Qduino Mini
RFM69
8×8 flexible matrix
LiPo battery

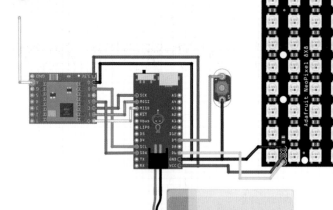

1200mAh 3.7V

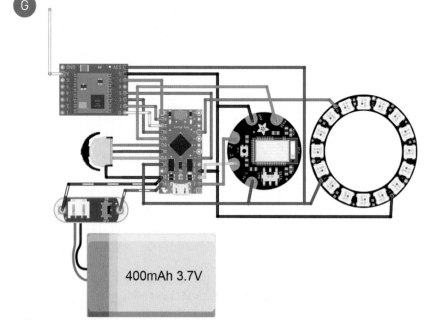

400mAh 3.7V

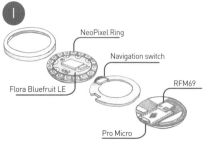

NeoPixel Ring
Navigation switch
Flora Bluefruit LE
RFM69
Pro Micro

3. ATTACHING LED PANELS

First, I carefully wrapped each LED panel around the back of the heel to test fit. To reduce bulk, I replaced the pre-soldered connector with silicone stranded wire. To disguise my wiring against the heel, I used black wires, marked with color-coded heat-shrink at the ends to keep track of them.

I used E6000 adhesive to attach the panels to the back of each heel with the wires running to the top corner of the panel near the inside of the arch. To hold the panels in place, I used small clamps and let the adhesive cure for 24 hours (Figure D).

4. HARDWARE IN HEELS

I chose to control each LED panel with a Qduino Mini because of its built-in power switch, micro-USB connector, and LiPo battery charging circuit — this minimizes the amount of hardware connections and space needed. The Qduino nests nicely in the arch of the shoe, with the switch and USB port facing to the side, so you can easily access the switch to turn the project on (Figure E).

NOTE: To save space, I used a transceiver module instead of a breakout board. Fritzing only has the breakout board artwork available, so these diagrams vary slightly from the build.

To test the RFM69 transceiver, I first prototyped the circuit on a breadboard. After adjusting the code and confirming functionality, I soldered all the necessary connections and glued the parts under the arch of the shoe. I ran the antenna of the RFM69 down the instep, bending as needed to avoid running underneath the toe and being stepped on. See Figure F for a full circuit diagram.

NOTE: I also added a LilyPad Button to the circuit for a future option of manual color/mode selection so the heels won't need to rely on wireless connection to the bracelet to function.

After testing the project, I covered the hardware and battery with gaffer's tape for some light protection from dirt and wear. So far the shoes have held up after a few nights out, but for long-term durability, I'll be prototyping a cover to fit over the electronics, probably 3D printed. Gaffer's tape was also used along the edges of the panels to disguise the seams and keep dirt from building up beneath them.

Angela Sheehan, Fritzing, Geoff Decker – Hidden Vision Photography

BUILD, HACK, & MAKE...

6 ISSUES FOR ONLY $39.99!

SAVE 33% OFF THE COVER PRICE

AUTONOMOUS LASER CAT TOY // RASPBERRY PINOCULARS

Make:

15 Fun DIY Projects

All Hands on Boards

Our Guide to 80+ Microcontrollers

Adafruit's Lunar Fried

Quadruped Robotics
Simple LED Copter
Stencil Spray Painting
Arduino Guitar Pedal

Vol. 57 June/July 2017

makezine.com | We Are All Makers

NAME
(please print)

ADDRESS/APT.

CITY/STATE/ZIP

COUNTRY

EMAIL ADDRESS (required for order confirmation and access to digital edition)

☐ Payment Enclosed   ☐ Bill Me Later

FOR FASTER SERVICE, GO TO: MAKEZINE.COM/ORDER

INCLUDES DIGITAL EDITION

B81NS2

5. PLANNING THE BRACELET

The bracelet was the most fun and challenging part of this build. After prototyping on a breadboard, I set out to make a compact wearable controller built into a watch-like module on my wrist.

To save on space, I used a Pro Micro instead of a Qduino and moved the battery connection to the band of the bracelet. For Bluetooth functionality, I chose the Flora Bluefruit LE module because its round shape nests nicely within a 16-LED NeoPixel ring. This creates a watch-like face for the bracelet and has a "statement piece" feel. An RFM69 transceiver broadcasts to the heels, and I included a 3-way navigation switch for an easy way to choose modes in future iterations. See the full wiring diagram in Figure ⓖ.

6. BUILDING THE BRACELET

I prototyped and laser-cut an acrylic casing for all the electronics (except the battery). A gold pendant found at a craft store created a cover for the casing (Figure ⓗ).

The first hardware layer houses the NeoPixel ring and Flora Bluefruit. The middle holds the navigation switch and an acrylic spacer with an open center to accommodate the wiring from the bottom layer. The base houses the RFM69 stacked on the Pro Micro, with engraved channels for wiring and the RFM69's antenna (Figure ⓘ). A piece of electrical tape over the Pro Micro protects it from any shifting that may occur with the RFM69 board above it.

Soldering everything together to fit was a bit of a puzzle, and cutting the wires as precisely as possible to reduce bulk inside the casing was a fun challenge.

Once I had it all assembled, the layers were reinforced with hot glue to avoid shifting during wear (Figure ⓙ).

7. FINISHING TOUCHES

To create a bracelet, the module is attached to black velvet ribbon with gold magnetic clasps.

I also glued a piece of clear acrylic inside the gold ring to protect the hardware, and I painted the white plastic of the LEDs black to minimize the contrast.

Finally, the battery powering the project is attached to a LilyPad Simple Power board near the clasp and tucked into a pocket under the band (Figure ⓚ).

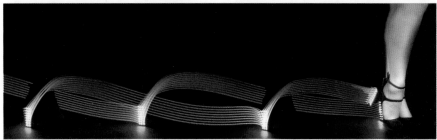

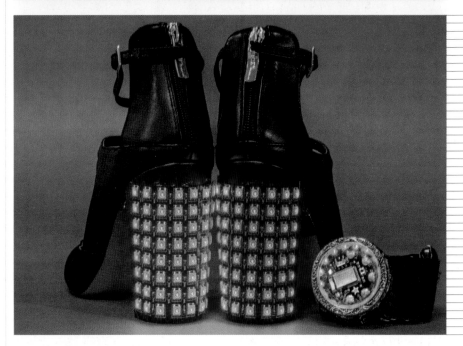

NEXT LEVEL KICKS

The first version of the Insta-Hue Heels uses a color wipe to update the color of the LED matrix from top to bottom. Even this simple animation is a crowd pleaser!

With a little extra programming, you can set your matrices to display complex animations and modes, chosen with a menu on the bracelet controlled by the navigation switch. I plan on adding some sparkling patterns, mirrored animations across the left and right heels, and other fun display options.

You can find updates to the functionality and code on my website along with more build photos and video demonstrations.

I'd love to see what you'll do with these heels! ◉

[+] Code and templates available at: github.com/GellaCraft/InstaHueHeels

[+] See more behind-the-scenes build photos and video at gellacraft.com/instahueheels

[+] SparkFun RFM69 hookup guide: learn.sparkfun.com/tutorials/rfm69hcw-hookup-guide

[+] Adafruit Flora Bluefruit LE guide: learn.adafruit.com/adafruit-flora-bluefruit-le

[+] Adafruit Bluefruit LE Connect guide: learn.adafruit.com/bluefruit-le-connect-for-ios

Hella Fast
Holograms

Use a laser and special film to create your own instant 3D images

Written by John Iovine

TIME REQUIRED:
1–2 Hours

DIFFICULTY:
Intermediate

COST:
$80–$100

MATERIALS

- » **Laser diode, 4mW–5mW** such as Model VM65003 from midwest-laser.com. I recommend purchasing from Midwest Laser or Litiholo, as these are proven hologram-producing laser diodes.
- » **Instant hologram film plates, 2"×3"** Litiholo #C-RT20, from litiholo.com
- » **LED safelight**
- » **Scrap of carpet, foam, or towel, 12"×12"**
- » **Ferrous metal plate, 12"×12"×⅛"** If your table is magnetic, you don't need this.
- » **Paint, flat black**
- » **Inner tube, 8"–12" dia.** for a small tire
- » **Cardboard squares, 3"×2" (2) one black and one white**
- » **Bar magnets, ⅜"×⅜"×1" (4)**
- » **Rectangular magnet, 0.9"×0.9"×0.4"**
- » **Steel plate, 2½"×1"×1/16"**
- » **Quiet and dark area**

TOOLS

- » **Table**
- » **Binder clips, medium size**
- » **Hot glue gun**

JOHN IOVINE is a science and electronics tinkerer and author who owns and operates Images SI Inc., a small science company. He resides in Staten Island, New York, with his wife and two children, their dog, Nigel, and their cat, Squeaks.

Holograms are fun to view, and recently they've become a lot more fun to make. Previously, shooting white-light reflection holograms — the kind you can view under ordinary light — could only be done using high-resolution holographic film that required development using chemicals. Not anymore. Now there's instant holographic film that doesn't require development, chemical or otherwise. Couple this film with the availability of inexpensive laser diodes, and it has never been easier to create holograms — it's really a shoot-and-see process!

WHAT ARE HOLOGRAMS?

Holograms are true three-dimensional pictures that capture the depth of a scene. True 3D pictures allow you to angle the hologram horizontally or vertically to get a different view of the subject, unlike standard 2D pictures which just create a foreshortened view of the same flat image when angled.

Holography doesn't record an image onto film the same way photography does. Instead, holography records the *interference pattern* of light, generated from a *reference beam* and reflected light from the subject (called the *object beam*). The light source required must be monochromatic (a single light frequency) and coherent (wavelengths in phase). A laser fits the bill.

WHY 5 MILLIWATTS?

A quick search for 650nm "focusable" laser diodes will show availability of much higher-powered laser diodes from Amazon and eBay, from 20mW to 200mW. You can use these, but I suggest a 4mW–5mW diode because they're safer — beaming yourself in the eye with one is less harmful.

MODIFYING THE LASER DIODE

You want to purchase a focusable laser diode (Figure Ⓐ) because the "focusable" lens is typically removable. This lens focuses the laser light into a beam, which produces a dot (Figure Ⓑ), but we want the laser light from the diode to spread out evenly, covering our holographic plate and the object we're shooting behind the plate (Figure Ⓒ). To make it so, remove the lens from the laser diode (Figure Ⓓ).

MAKING AN ISOLATION TABLE

When shooting holograms, vibration must be eliminated as much as possible. Vibrations too subtle for us to feel, such as those caused by music or ceiling fans, can prevent a hologram from forming. Holographers shoot holograms on an isolation table. As its name implies, an isolation table "isolates" the holographic setup from vibration.

A simple isolation table I use is illustrated in Figure Ⓔ. It consists of three components: a piece of carpet (or soft foam, or thick cloth like a towel), a small 12"-diameter inner tube, and a 12"×12"×⅛" thick ferrous metal plate.

Fill the inner tube with just enough air for it to be full but still soft and easy to squeeze. The 12"×12" metal plate needs to be thick enough to support itself and a few lightweight components without flexing or bending, and it needs to be a ferrous metal so that our magnetic mount will stick to it.

Paint the steel tabletop flat black. This will help cut down on unwanted laser light reflection to improve the quality of your holograms.

You'll want to set all this up in a place that is dark and quiet.

Ⓐ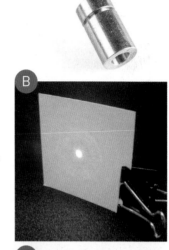

Ⓑ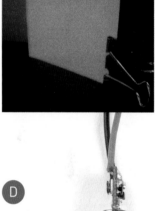

Ⓒ

Ⓓ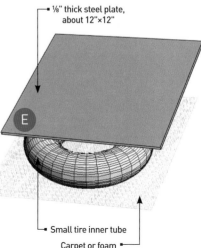

ISOLATION TABLE DIAGRAM

⅛" thick steel plate, about 12"×12"

Ⓔ

Small tire inner tube

Carpet or foam

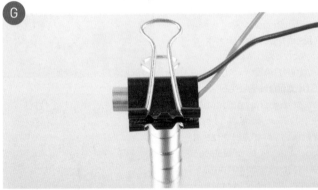

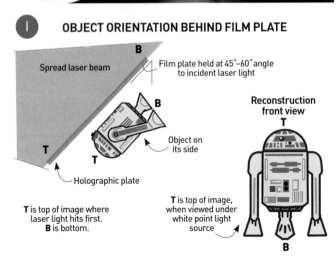

OBJECT ORIENTATION BEHIND FILM PLATE

Spread laser beam

Film plate held at 45°–60° angle
to incident laser light

Object on
its side

Holographic plate

Reconstruction
front view

T is top of image where
laser light hits first.
B is bottom.

T is top of image,
when viewed under
white point light
source

MOUNTING THE LASER

For our basic single-beam setup you'll only need a few mounts and components.

You can create mounts using small permanent magnets, the metal isolation table surface, and small steel plates or binder clips. The magnets can either be ceramic or high-strength neodymium.

I typically hot-glue the object I'm going to create a hologram of onto a small steel plate, 1"×2½"×1⁄16". When you secure the plate or binder clip to the steel isolation table using a magnet (Figure F), it allows for easy repositioning.

Binder clips with magnets are also useful for holding the holographic film, white setup card, black shutter card, and laser diode. The entire diode fits into a medium sized binder clip, which also serves as a heat sink (Figure G).

MOUNTING THE FILM

To secure the glass holographic film plate (Figure H) to the table, you'll place the film in a medium binder clip, secured to the table with a magnet to prevent vibration.

SAFELIGHT

When handling the holographic film, you'll darken the room and use a *safelight* to see what you're doing. The film is sensitive to red, green, and blue light. Liti Holographics suggests using a single-diode blue LED safelight, which can be fabricated from a battery, series resistor, and a darkroom-safe LED (I used a green one because I have them from my other holography work). Use your safelight as little as possible to prevent fogging your film.

POSITIONING YOUR MODEL

For your first model, choose something rigid and hard that won't bend, droop, or move during exposure. If the object is dark, paint it white. Ideally the complete object will fit behind the 2"×3" film plate. For beginners I recommend a small, light-colored seashell.

To create the brightest viewable hologram, position the model close to the film plate without touching (Figure H).

SETTING UP YOUR HOLOGRAPHY TABLE

I use a side-lighted, single-beam holographic setup. To understand what this means, trace the spread beam from the laser diode to the holographic plate (Figure I).

With this side-lighting setup, the point where the spread beam first touches the plate is labeled *T* because it will be the top of the holographic image when viewed later under a white light. Therefore the top of your model when placed behind the plate should be oriented to the *T* side of the plate. Figure J shows a diagram of the entire table, and Figure K shows it from a different angle, with the object, film, and laser diode.

To make sure your laser light is correctly oriented, test it by placing a 2"×3" piece of white cardboard where you will place the film. Use a binder clip secured with a magnet to keep it in position. Direct the spread laser light onto the white card, making sure the light strikes the card at a 45°–60° angle. Position the diode so that laser light fills the white card as evenly as possible.

Next, position the model behind the white card. Remove the white card, leaving behind the binder clip and magnet. With the white card removed, the laser light will be illuminating the object. Look at the object from the laser side. This will be the view of your finished hologram.

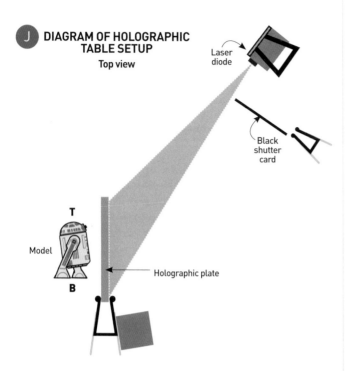

J DIAGRAM OF HOLOGRAPHIC
TABLE SETUP
Top view

Laser diode

Black shutter card

Model

T

B

Holographic plate

SHUTTER, EXPOSURE, HOLOGRAM!

Use a 2½"×2½" piece of matte black cardboard as a shutter, propped up with a binder clip. Don't use magnets on this binder clip, because you'll need to lift the shutter card easily.

Block the laser light using the shutter card. Turn all the lights off and remove a film plate from its light-tight box. Hold the plate by its edges, turn on your safelight, and place the film in the binder clip that was holding the white card. To ensure maximum stability, make sure the bottom edge of the plate and the binder clip are flush with the surface of the table.

To expose the holographic film, lift the card off the table to unblock the laser light. The exposure takes 10 minutes, and then you can place the shutter back down in front of the laser.

The exact exposure time varies with the diode's power output, and how wide the laser light is spread. Litiholo recommends using a 5–10 minute exposure time using a 5mW laser diode, and that it's better to overexpose than underexpose.

To develop the film, just turn on the room lights.

ADMIRE YOUR WORK

Now it's time to take a look at your hologram. The hologram you shot is a *white light reflection hologram* (viewable in standard white light). You can obtain the sharpest image by using a point light source above, and a black background behind the hologram. The sun, tungsten halogen spots, and bright white LEDs work well. Incandescent lamps can be used, but the image quality will suffer.

To view the hologram, the point light source must illuminate the hologram at the same angle as the laser beam did during exposure (Figure **L**). So you may have to rotate the hologram a couple of times and vary the angle until the light strikes the plate correctly. If you're holding the holographic plate upside down (in reference to the angle of the laser light used to make the hologram) or sideways, you may not see any image at all.

To improve the appearance of the image, place something black behind the hologram.

TROUBLESHOOTING

While shooting holograms is easy, shooting *good* holograms takes a little practice. I have over 20 years experience, and have written numerous articles and two books on the subject. Here are some common complications that can arise when shooting holograms.

No Image — Ensure there is nothing that can vibrate the table during exposure: music, a running fan, drafts, etc. This is the primary reason holograms don't form. You can also try a longer exposure time; again, it's better to overexpose than underexpose.

Laser Diode — It's possible that the particular diode you purchased may not produce a hologram. Two manufacturers sell laser diodes that have already been tested to produce holograms: Midwest Laser and Litiholo.

Faint Image — Can be caused by stray light fogging the film. Make sure there are no other light sources besides the laser. Use your safelight as little as possible, and do not hold it close to the holographic film.

When viewing, make sure the holographic film plate is oriented properly with reference to the point light source.

GOING FURTHER

After you've practiced your holography skills using this single-laser setup, you might be interested in trying full-color holograms. Litiholo has developed a full-color kit that uses the same instant film but requires three lasers (red, blue, and green) to shoot the hologram (Figure **M**). The process is a little more complicated but the results are even more lifelike. ◔

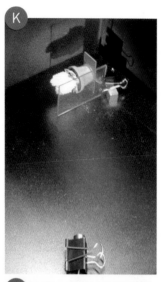

K

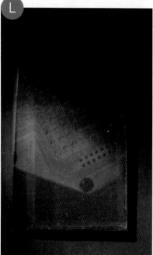

L

TIP: During the long exposure time, your laser diode may *mode hop*. This creates an interesting echo effect in the resulting hologram, almost like the image was shot multiple times. Hard to explain, but if you see it, you'll say, "Oh yeah, that's it."

Mode hopping is when the laser diode changes its light output frequency slightly, due to heat and current fluctuations in the crystal cavity and such. The only simple way to reduce mode hopping is to under-power the laser diode. So if the diode is rated at 5V, give it about 4.5 volts. It helps, but it's not a cure.

M

Get Your Motor Runnin'

Beginners, start using electric motors for real robots! **Written by Gordon McComb**

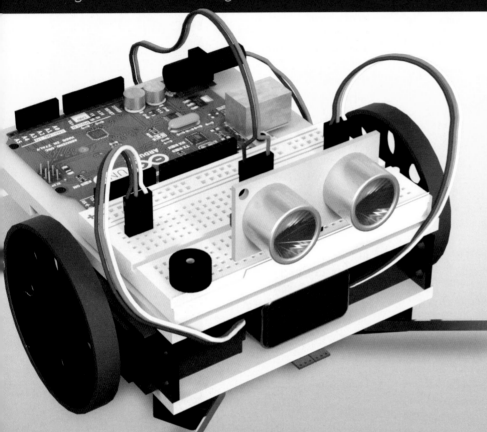

Here's a preview of the "electric motors" section of our new book for beginners, How to Make a Robot — yes, a real robot that follows your commands! We call it **Make:Bot**. Build it from scratch using common electronic parts plus tools and materials you can find at craft and hardware stores. The inexpensive, easy-to-use, programmable Arduino microcontroller makes it smart.

Make:Bot is really five robots in one, which show five major functions of robots:

» **Tai Chi Bot** replicates a set pattern of preprogrammed movements.

» **Touch-and-Go Bot** uses "whiskers" to navigate its environment.

» **Bat Bot** sends out sound waves to detect and avoid obstacles.

» **Remote Bot** lets you use a TV remote to control it.

» **Line Bot** follows a line drawn on poster board.

Basic soldering and Arduino programming skills are recommended (it's easy). Otherwise, you don't need any prior experience to create your own Make:Bot. There's even a kit!

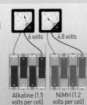

GORDON MCCOMB, author of more than 65 books and thousands of magazine articles, has been called "the godfather of hobby robotics" by none other than *Make:* magazine.

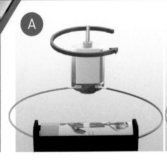

Turns about 4x faster

CAUTION: Do not try to power this type of ordinary DC motor directly from the Arduino, or you could wreck your Arduino!

6 volts   4.8 volts

Alkaline (1.5 volts per cell)   NiMH (1.2 volts per cell)

AA BATTERIES AND VOLTAGE

Alkaline AA batteries produce **1.5 volts** per cell. A set of four batteries wired in series make **6V**.

Rechargeable AA batteries, such as nickel metal hydride, produce **1.2 volts** per cell, or **4.8V** for four. When using rechargeable batteries, the reduction in voltage causes the robot motors to run slightly slower.

POWERING ELECTRIC MOTORS

The ability to travel from one place to another is an essential characteristic for many robots. This movement is often provided by **electric motors**. Applying **voltage** to a motor makes it spin, and this moves the robot.

A **battery** or **battery pack** provides power to the motor. To change the direction of the motor, simply reverse the **+** and **−** terminals from the battery (Figure Ⓐ).

Two important aspects of motor operation are speed and torque. **Speed** is the rotational velocity of the motor, typically specified in revolutions per minute (RPM). **Torque** is the amount of force the motor exerts.

The best way to increase the speed and torque of a motor is to add more batteries in series. Put four batteries in a four-AA holder, making sure positive and negative terminals are facing the proper direction.

Connect the motor to the battery holder. What happens?

The motor spins about four times faster because it's now getting 4x the voltage (Figure **B**). The motor also has **more torque.** You need the extra torque to move your robot around!

SERVO MOTORS

The Make:Bot uses a special type of motor, a **servo motor** that's similar to those in radio control (R/C) models and toys (Figure **C**). But unlike regular R/C servo motors, which are limited to turning through a 90° (or so) arc, hobby servos made for robotics may turn **continuously** in either direction.

The operation of the motor is set by sending a series of pulses of varying durations, from 1 to 2 **milliseconds** (Figure **D**). This control signal can be easily created using an Arduino microcontroller, the brains of our robot.

CAUTION: Never *ever* reverse the polarity of the battery connections to the servo, or damage may result!

ARDUINO + SERVOS

An **Arduino microcontroller** (Figure **E**) operates all the functions of the Make:Bot. Software running inside the Arduino determines what the robot is supposed to do. **Input/output (or I/O) pins** on the Arduino provide connections to your robot's servo motors and other hardware. Separate **power pins** allow you to provide electrical juice to various components you add.

But you'll power your servos separately, so they won't be a drag on the microcontroller. Make:Bot uses two battery sources: a 9V battery that runs the Arduino, and 4xAA batteries (4.8 or 6 volts) for the servos.

This schematic diagram (Figure **F**) shows how the robot's servos are wired up.

GET THE KIT!

Start building your Make:Bot! Get a kit of all the electronic parts plus the *How to Make a Robot* handbook at makershed.com/products/make-how-to-make-a-robot.

Or if you prefer to learn on your digital device, grab the kit and look for our new online learning workshop at makershare.com/learn. ✎

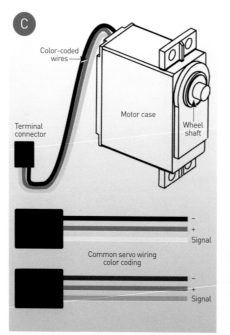

Color-coded wires

Motor case

Wheel shaft

Terminal connector

−
+
Signal

−
+
Signal

Common servo wiring color coding

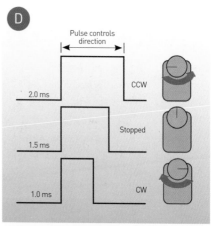

Pulse controls direction

2.0 ms — CCW

1.5 ms — Stopped

1.0 ms — CW

Core benefit of hobby servos: They have built-in driver circuitry, so they can be controlled directly from the Arduino's I/O pins.

It's also easy to attach wheels to servos, and to mount them to your robot.

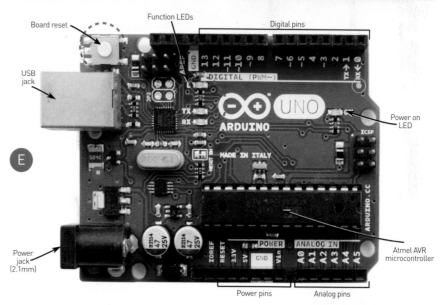

Board reset

Function LEDs

Digital pins

USB jack

Power on LED

Power jack (2.1mm)

Atmel AVR microcontroller

Power pins

Analog pins

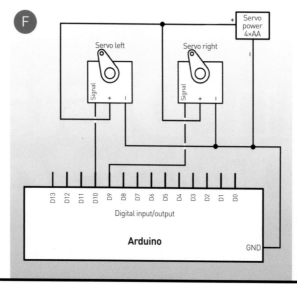

Servo power 4×AA

Servo left

Servo right

Signal

Digital input/output

Arduino

GND

CAN I JUST USE THE 9V BATTERY TO RUN THE MAKE:BOT?
No. The 9V battery does not provide enough current to operate the servomotors.
HOW ABOUT JUST ONE BATTERY PACK?
Not recommended. Arduino requires at least 7 volts, but most servos are made for just 4.8V–6V. And, it's not advisable to run servos from the Arduino's 5V regulated supply.

Gordon McComb

Mini **Distance**
Detectors

Two tiny IR sensors let your project locate nearby objects

Raul Arias, vecteezy.com, Forrest M. Mims III

TIME REQUIRED:
1–2 Hours

DIFFICULTY:
Easy

COST:
$15–$50

MATERIALS
» **Near-IR distance detection module, triangulation type** Pololu #2474, pololu.com
» **Batteries, AAA (2)**
» **Battery holder, 2xAAA**
» **Wire, fine gauge** such as 30AWG wrapping wire

FOR OPTIONAL VOLTAGE-CONTROLLED OSCILLATOR:
» **TLC555 timer IC chip**
» **Mini loudspeaker, 8Ω**
» **Resistors: 68Ω (1) and 680Ω (1)**
» **Capacitors: 2.2µF (1) and 47µF (1)**

TOOLS
» **Voltmeter**
» **Wire wrap tool and wire cutters**
» **Soldering iron and solder**

FORREST M. MIMS III (forrestmims.org), an amateur scientist and Rolex Award winner, was named by *Discover* magazine as one of the "50 Best Brains in Science." His books have sold more than 7 million copies.

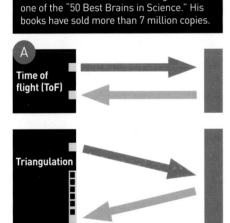

A Time of flight (ToF)

Triangulation

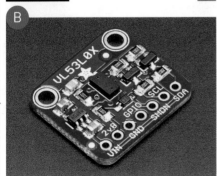

Adafruit breakout board with STM's ToF distance module.

While designing miniature travel aids for the blind years ago, I used a brute-force method of detecting objects. A string of high-current pulses was applied to a high-power near-infrared LED. The pulsating IR beam, reflected from objects up to 12 feet away, was detected by a 1-inch square silicon solar cell, amplified, and sent to a small earphone. The phone emitted an audible tone at the frequency of the pulsating IR. The volume of the tone provided a clue to the distance to the object.

The problem with this method was that distant targets with good IR reflectance, such as white objects or green tree leaves, produced a similar volume as nearby dark objects with much poorer reflectance. So, I experimented with a triangulation method that provided a much better indication of distance. This handheld device featured a photodiode installed behind a lens in a tube that could be slightly rotated when pushed down. When the travel aid emitted a tone that indicated an object was present, the blind person pressed a button that pushed down the photodiode lens tube. The position of the lens tube that provided the loudest tone provided a rough indication of the distance to the object.

While this method worked, it was Stone Age compared to today's off-the-shelf distance detectors. Portable long-range laser distance detectors that once cost $20,000 can be purchased for only a few hundred dollars. Short-range detectors include both ultrasonic and active near-IR systems that indicate the distance to objects up to a few yards away. Some of these short-range sensors are available on breakout boards, ready to connect to an Arduino or similar controller, for under $20. Others provide an analog voltage output that's proportional to distance. These devices can provide distance sensing for robotic devices, drones, and toys. They can also be used to switch doorbells, devices, and appliances on and off when a person approaches.

The engineering of these distance detectors is ready to go. All that's necessary is to connect them to a relevant circuit or controller and apply power. We'll get started by examining *time-of-flight* near-IR distance sensors. Then we'll look at a simpler family of near-IR distance detectors that employ *triangulation* to estimate distance. The two methods are diagrammed in Figure A.

TIME-OF-FLIGHT DISTANCE DETECTORS
The most sophisticated near-IR distance sensor modules determine distance to an object by measuring the time required for a flash of near-IR from a laser to be reflected from an object and back to the module's detector. This is not simple, for light travels 1 foot in 1 nanosecond. Yet STMicroelectronics has developed tiny, time-of-flight (ToF) distance detector modules that measure only 2.8mm×4.8mm, about the size of a rice grain. These remarkable devices include a microprocessor, range circuitry, laser, and an avalanche photodiode that can detect single photons. The distances they measure are unaffected by the target's color and texture. While transparent or mirrorlike surfaces might not be detected as well as targets having a diffuse surface, the distance will be accurate unless multiple reflections occur.

While details of ST's FlightSense technology are apparently proprietary, a key hint is given in their patent EP 2728373 A1, "Improvements in time of flight pixel circuits":

"Said measurement circuit is operable to discharge said capacitance at a known rate over a discharge time period, the length of said discharge time period being determined by the time of said detection of said photon incident on said SPAD [Single Photon Avalanche Diode], such that the final amount of charge on said capacitance is an analogue representation of the time of flight of said photon."

While the use of a SPAD is impressive, the most interesting aspect of this disclosure is that a high-speed clock is not used to measure the time of flight. Instead, the charge on a capacitor is directly proportional to the time of flight, which can be tenths of a nanosecond.

ST's VL53L0X Time-of-Flight Distance Sensor has a detection range from around 1 inch to 3 feet or more. Breakout boards with the VL53L0X and associated circuitry are available from Adafruit, Pololu, and other vendors for less than $15. Other modules with a closer range are also available. Adafruit's description of their tiny breakout board (Figure B) as a "micro-lidar" is certainly appropriate for this amazing system.

Pololu near-IR distance module.

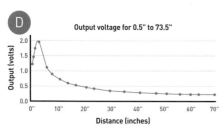

Output voltage for 0.5" to 73.5"

Voltage vs. distance for a Pololu near-IR distance module.

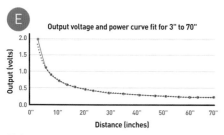

Output voltage and power curve fit for 3" to 70"

Voltage vs. distance for a Pololu near-IR distance module, with first three points deleted.

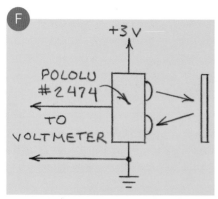

Basic Pololu near-IR distance module test circuit.

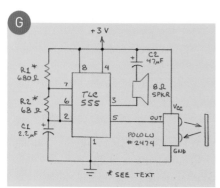

Pololu near-IR distance module plus TLC555 voltage-controlled oscillator.

TRIANGULATION DISTANCE DETECTORS

Triangulation distance detection systems employ a much simpler strategy than ToF systems. A typical triangulation sensor employs a forward-facing, near-IR LED adjacent to a row of photodiodes behind a lens. The photodiode that receives the highest level of near-IR reflected from an object indicates the distance to the object.

A family of near-IR distance measuring devices with an analog output is made by Sharp. Included is the Sharp GP2Y0A60SZLF, a long number for a part about the size of a 14-pin DIP and slightly thicker. This analog device emits a voltage that's proportional to the distance to a detected target. It requires 5 resistors and 8 capacitors, all of which are installed together with the distance module on a circuit board available from Pololu (pololu.com/product/2474). This assembly measures only ⅜"×1¼" (Figure C). The most impressive feature of this module is that the source and detector lenses are only ¼" in diameter. Also impressive is that the module works well in sunlight.

The Sharp data sheet includes considerable information about using this module. To verify the data sheet's plot of output voltage versus distance, I mounted a Pololu module on my workbench and pointed it at a movable sheet of white plastic. The module's output voltages for a range of distances from the target were entered into a spreadsheet. Figure D shows the resulting chart, which closely resembles a chart in Sharp's data sheet.

Note that the chart shows a sharp change in the output voltage when the target is less than a few inches away. In Figure E, the closest three measurements have been removed, and the result is a nearly perfect power curve with an R^2 of nearly 1.

BUILD A TEST CIRCUIT

Both ToF and triangulation near-IR distance measuring modules are supported by lengthy guidelines that you should read carefully to understand their operation and limitations. Before digging into these, you can use Pololu's #2474 triangulation distance measuring module with an analog output for a very quick, hands-on introduction to near-IR distance measuring.

As shown in Figure F, all you need is the module, a 3-volt battery (two AAA cells in a holder), and a voltmeter. You can use wrapping wire to quickly connect the battery to the module by first soldering the included 1×4, 0.1" right-angle header to the module.

Figure G shows a circuit that converts the analog voltage from the Sharp GP2Y0A60SZLF module into a tone with a frequency that varies with distance. The module's analog output is connected to a TLC555 timer configured as a voltage-controlled oscillator (VCO). The TLC555 is a CMOS, low-voltage version of the standard 555. The VCO drives a miniature 8-ohm speaker. In operation, nearby objects give a higher-frequency tone than objects farther away. I built the circuit on a breadboard and then made a miniature version (Figure H) that fits inside an Altoids flip-top tin together with the mini speaker, two AAA cells, and Sharp module.

This ultra-simple setup can detect a ⅛" fence wire 2 feet away, and a 3½" wood post 10 feet away. These are maximum ranges; it performs best at close range, since unfortunately there is no obvious change in the frequency of the tone for objects more than a few feet away.

CHOOSING THE RIGHT SENSOR

For basic object detection, a triangulation mode, near-IR distance detector works well, especially for short distances. For more precise measurements, a ToF "micro lidar" detector circuit is best, but a microcontroller is required. Ultrasonic detectors are also a strong possibility (see Make: *Volume 61, page 56, "Robot-Ready Radar"*) — I'll cover these in a future column. ●

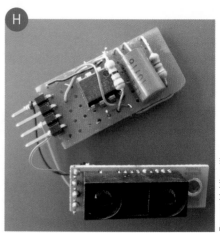

Photo of near-IR distance module and voltage-controlled oscillator circuit.

Written and photographed by Glen Scott

TIME REQUIRED:
1.5 Hours

DIFFICULTY:
Intermediate

COST:
$16

MATERIALS
» **Steel square tube, 2½"×1"×¹⁄₁₆"** from a hardware store
» **Hardwood scraps**
» **Threaded rod, ⁵⁄₁₆"-18 thread, 7"–10" length**
» **Star knob with ⁵⁄₁₆"-18 threaded post**
» **Threaded inserts, ⁵⁄₁₆"-18 internal thread (2)** for hardwood. I used a jig kit, Amazon #B01DRGF1IE, with the knob and inserts.
» **Washers, large (2)** for top and bottom of swivel pad
» **Washers, small (2)** for inside swivel pad
» **Felt padding**

TOOLS
» **Hacksaw and small wood saw**
» **Center punch**
» **Mallet**
» **Drill and bits, Forstner bit**
» **Hole saw**
» **File**
» **Pliers**
» **Adhesive glue** such as epoxy or Gorilla Glue
» **Clamps**
» **Lathe and angle grinder** (optional)

GLEN SCOTT is an experienced DIYer. He makes and shares doable projects in step-by-step videos, mostly done with simple tools.

Save Face

Hold wood trim to the face of a workpiece with this clever bar clamp attachment

E ver had trouble holding a piece of wood trim to the face of another workpiece? I'm sure many of us have been there. This clamp attachment is a cool way to fix the issue. It's designed to fit a typical bar clamp — just attach it and tighten the knob.

1. CUT AND DRILL THE TUBE
Use a center punch to mark the square tube about ½" from the end, then drill a ½" hole (Figure A). Keep the bit straight, you want to go all the way through the tube and exit the other side.

About ¾" from the other end, notch out a slot for your clamp's bar to rest in. Use a hacksaw to cut out the slot (Figure B), then file it down to clean out the rough edges.

2. ADD WOOD AND THREADED INSERTS
Cut a small block of wood about 2½"×1"×1" and drive it into the metal tube (Figure C). You can add a bit of epoxy or Gorilla Glue to secure it. Then cut out the wood where the metal is notched out (Figure D).

At your ½" hole, drill again (Figure E) to install a threaded insert (Figure F). This insert will play the key role in holding down the workpiece.

In the opposite end, install the second threaded insert. This one will lock the attachment to your clamp bar (Figure G).

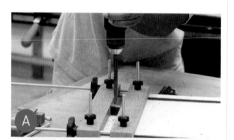

A

B

3. MAKE THE CLAMP SCREW

Carefully cut a groove into the threaded rod as shown (Figure **H**). I used my grinder locked in a vise.

To make a retaining ring, choose a washer with a small enough hole to keep it in the groove. Cut it (Figure **I**), pry it open, place it in the groove, and then close the washer (Figure **J**).

4. MAKE A HANDLE

I made a hardwood handle on the lathe (Figure **K**), drilled a hole into it, and then used a piece of threaded rod to cut threads into the handle (Figure **L**).

5. MAKE THE SWIVEL PAD

In a wood block 1" or thicker, use a Forstner bit just bigger than your small washer to remove about ½" of material (Figure **M**). Next, take a hole saw the size of your large washer (Figure **N**) and drill out the entire hole. Remove the plug and cut it down to 1" in length (Figure **O**). This is your pad.

Slip the rod through the wooden pad, and glue the small retaining washer inside the lip of the pad. Put the other small washer at the end of the threaded rod — its hole must be smaller than the rod, so the rod cannot exit (Figure **P**).

Now take a large washer and glue it to the bottom of the wooden pad (Figure **Q**). (It's a good idea to attach a second large washer on the topside of the pad — ideally, you'll want to sandwich the wooden pad in between.) Clamp the pad and washers together and let the glue set.

6. FINAL ASSEMBLY

Finally, add a 1" felt pad to the washer to protect your workpiece (Figure **R**), thread the rod through the attachment bar, then glue and screw the handle onto the rod. Screw the star knob into the end.

USE IT

The attachment is a great way to hold down your work when you need to attach one piece to the face of another, whether it's horizontal (Figure **S**) or vertical (Figure **T**). I hope you find it to be useful in your workshop. ✏

[+] Watch the build and get more tips at
youtu.be/Is67ZeGAdqo

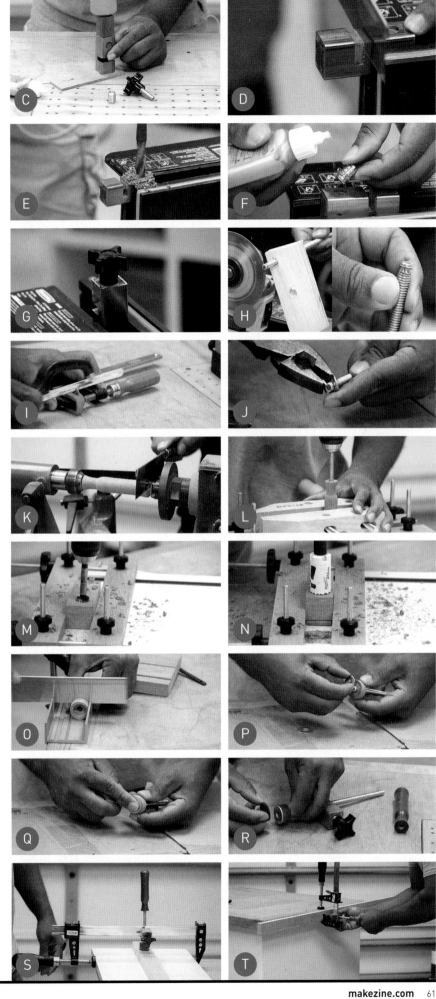

A Logical Oscillator

Skip the 555 and create audio frequencies using an inverter chip instead

Written by Charles Platt

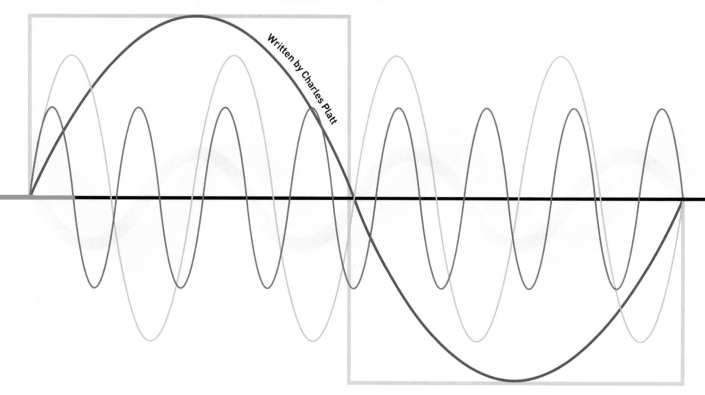

IF you look inside any electronic device that flashes a light or makes a noise, you'll find a circuit that oscillates. A 555 timer is the traditional building block to create oscillations — but I want to show you how to do it with fewer ancillary components at a fraction of the cost, using the world's simplest logic gate.

This logic gate is an *inverter*, represented by the symbol in Figure Ⓐ. Oscillating inverters can create sound effects for robots, video games, and much more. A chip containing six inverters retails for around 50 cents, and with two chips, you can generate the 12 semitones in a musical scale. Double the number of chips, and you have a two-octave synth. If you can salvage a keyboard from a discarded toy, the whole thing could cost around $10.

An inverter has only one input and one output. A high state on the input forces a low output, while a low input creates a high output. This isn't very interesting until you connect the output back to the input, as in Figure Ⓑ.

Now when a high input creates a low output, it circles around to create a low input, which causes a high output, which circles around to create a high input, forcing a low output — in a cycle that repeats about 10 million times per second.

The circuit is now oscillating. We just need to slow it down.

If you've read my book *Make: Electronics*, you know that a resistor in series with a capacitor can adjust the speed of an electronic signal. In Figure Ⓒ, the capacitor is wired to the inverter input, so the input can't flip instantly into a low or high state. It has to wait for the capacitor to charge or discharge through the resistor. The resistor and capacitor values we choose will set the frequency of the oscillations.

However, there's a problem. If the inverter is triggered by even a tiny excursion above or below the dividing line between high and low input states, small variations in temperature or power supply can make it behave erratically.

Fortunately a circuit called a Schmitt trigger is built into some logic chips. Any time you see the symbol in Figure Ⓓ, you know that a component contains this feature. It introduces a gap between the low and high input voltages.

A high Schmitt input in a 5V logic gate is considered 3V, a low input is 2V, and in the gap between them, the inverter ignores any small variations and maintains a constant output. The four snapshots in Figure Ⓔ show how this works.

An inverter with a Schmitt trigger is stable enough for simple applications, and is easily adjustable with a trimmer potentiometer, as in Figure Ⓕ. Just be sure to add a fixed resistor to protect the chip when the trimmer is turned down to zero.

Figure Ⓕ also shows that when you tap the inverter output, you get a square

TIME REQUIRED:
2 Hours

DIFFICULTY:
Easy

COST:
$15–$20

MATERIALS
» 9V battery or AC-DC adapter (9VDC or 12VDC), or adjustable power supply (up to 12VDC)
» Solderless breadboard, full-size
» Jumper wires, assorted
» Loudspeaker, 3" diameter, 8-ohm
» Enclosure of your choice for loudspeaker
» Trimmer potentiometers: 50kΩ (2), 500kΩ (2), and 2kΩ (1)
» Capacitors, ceramic, minimum 16VDC: 0.1µF (1), 0.33µF (4), and 10µF (1)
» Capacitors, electrolytic, minimum 16VDC: 220µF (1), and optionally 1,000µF (1) if needed to suppress hum
» Resistors: 1kΩ (2) and 100kΩ (2)
» Diodes, 1N4148 (2)
» 40106B CMOS hex inverter IC chip, through-hole, PDIP type
» LM386 amplifier IC chip

TOOLS
» Multimeter
» Wire strippers
» Pliers

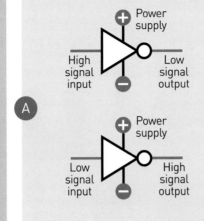

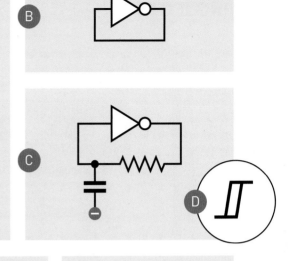

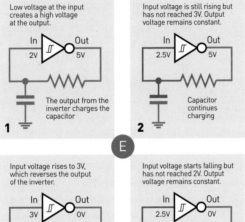

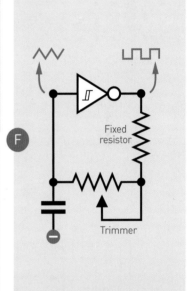

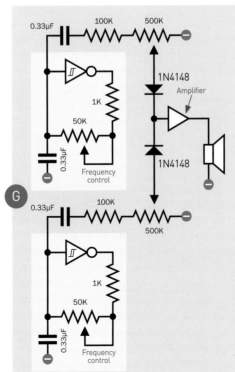

wave, because the output voltage flips from maximum to minimum almost instantly. If you tap the input, you get a sawtooth wave, because the input is controlled by the charging and discharging of the capacitor. At audio frequencies, a sawtooth wave sounds more mellow than a square wave.

For a synth, you would need one inverter for each note, all of them sharing an amplifier. The first step toward this goal is to see how two inverters can be coupled together, as in Figure G. Conventionally, you would use transistors or op-amps to mix the oscillator outputs, but "cheap, simple, and sounds okay" is my motto here. Also, the oscillators must be protected from even small fluctuations that feed back into the power supply. Therefore I used diodes to couple the oscillator outputs while allowing gain adjustment of one channel without affecting the other. Audiophiles may not like the look of this, but I think the sound is satisfactory.

CHARLES PLATT is the author of *Make: Electronics*, an introductory guide for all ages, its sequel *Make: More Electronics*, and the 3-volume *Encyclopedia of Electronic Components*. His new book, *Make: Tools*, is available now. makershed.com/platt

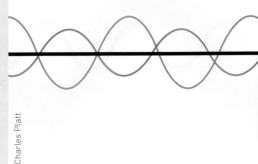

Charles Platt

A breadboarded pair of inverter-oscillators.

Charles Platt

Breadboarding this circuit is awkward because there isn't room for trimmers alongside the inverter chip. In Figure **H**, I redrew the schematic so that transferring it to a breadboard will be a bit easier. Figure **I** shows the finished result.

The LM386 chip is the amplifier. You adjust its gain between a factor of 20 and 200 using the 2K trimmer or a fixed resistor of your choice, with the 10µF capacitor. It drives a 3", 8-ohm speaker in a project box to reproduce the lower frequencies.

The inverter chip is a 40106B. It can tolerate a power supply up to 18V, and the LM386 can use 12V, so you can run the circuit directly off a 9V battery with no need for a voltage regulator. However, you should add a regulated power supply if you want the pitch of each note to be consistent.

Electrical noise is always an issue when using an amplifier. To suppress it, the 0.1µF capacitor shown near the inverter chip is essential. Add a 1,000µF capacitor across the power supply if you notice humming or buzzing sounds.

Tie any unused inverter inputs to the negative ground bus, but leave the outputs floating. Use ceramic capacitors throughout, except for the 220µF output capacitor and the optional 1,000µF, which are electrolytic. Capacitors should be rated for at least 16VDC.

You can have fun with this circuit. Substitute a much higher value for the 0.33µF timing capacitor in one oscillator, and it will create a vibrato effect in the other. Use a 10µF capacitor to bridge two inverters, and you'll hear a sound like a monster growling inside a cardboard box.

But what about that bargain-basement synthesizer? Just add more oscillators, each linked through a diode to the amplifier input. Adjust the pitch of each using a guitar-tuning app, and switch them using a keyboard. That's all there is to it. ◉

[+] For technical information about Schmitt triggers, search online for Application Report SCEA046, published by Texas Instruments. To learn more about logic oscillators, take a look at Application Note 118 published by Fairchild Semiconductor. You'll find there are a lot of possibilities.

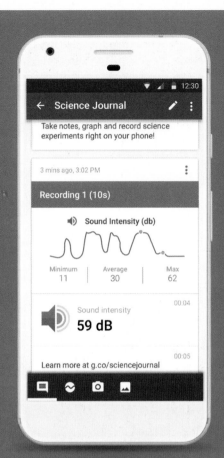

OpenBraille

Written by Carlos Campos

Build a braille embosser for a fraction of the cost of commercial machines

A

B

C

CARLOS CAMPOS studied physics and mathematics at McGill University and is finishing a second degree at Polytechnique Montréal as a mechanical engineer. His biggest dream is to become an inventor, develop products, and start companies.

This project started when I met Alvino. He is originally from the Bahamas and he was born unable to see. Today Alvino has migrated to Canada where access to braille embossers is subsidized by the government, but still they're expensive, and repairs are costly. I started to think about building a machine for him to read music as he plays the accordion.

The initial idea led me to think about a system of pins and a rotary encoder to build a new kind of embosser. Commercial machines emboss the paper by impacting it all at once. OpenBraille uses a physical encoder and a roller; this way, the embossing is done gradually, requiring less force, so we can easily use 3D-printed parts.

I started by designing and printing the encoder (Figure Ⓐ). As the embossing head moves back and forth across the paper, three pins punch the braille dots. These pins are actuated by a 3D-printed encoder wheel with three separate tracks, in a cam-and-follower arrangement, to form each braille letter correctly.

The three encoder pins, each made from a nail and a hex nut (Figure Ⓑ), are the only components that have to be machined; you can use a Dremel and Vise-Grips.

I then used the typical components of a 3D printer and pieces of plywood to build the rest of the machine (Figure Ⓒ).

After three weeks of working day and night, I managed to emboss my first card. When I presented it to Alvino, he was able to read the message: "Hola Alvino."

Because the braille printer is based on existing 3D printers, there was very little programming to do: I had to modify the firmware for the Arduino, and I wrote some Python code to translate the braille letters into G-code.

OpenBraille is a production of LaCasaLab, a homemade laboratory made by me and my roommate Christelle Fournier. Special thanks to Sensorica and Eco2Fest, and to David Pache who programmed the user interface. The project also won grand prize in Instructables' 2017 Arduino contest.

This isn't, by any means, a finished product. By making the machine an open source project, I'm hoping others will improve the design. If you're curious or you want to help out, feel free to follow our tutorial and help us build a community around OpenBraille. ◉

TIME REQUIRED:
A Weekend

DIFFICULTY:
Intermediate

COST:
$210–$300

MATERIALS
» **3D printed parts** Download the free 3D files at thingiverse.com/thing:2586738
» **Arduino Mega microcontroller board** such as Abra Electronics #A000067, abra-electronics.com
» **RAMPS 1.4 3D printer controller board** aka "shield" for Arduino Mega, Abra #3D-E-008
» **Stepper driver boards (3)** Abra #MOT-A4988
» **Endstops (2)** Abra #3D-E-002
» **Servomotor, micro size** Abra #FS90
» **Stepper motors, NEMA 17 size (2)** Abra #MOT-SM-17
» **Linear rod, 8mm dia., 400mm lengths (2)** Abra #MP-R-07
» **Rod clamps, 8mm (4)** Abra #MP-MRC-8
» **Leadscrew rods, M8×400mm (2)** Abra #3D-H-011
» **Pillow block bearings, 8mm (4)** Abra #MP-MPB-8
» **Linear bearings, 8mm (4)** Abra #MP-SBS-8
» **Shaft couplers, flexible (2)** Abra #1176-ADA
» **Screws, M3** such as Amazon #B01MRP19TB
» **Power supply, switching, 12VDC 10A 120W** Abra #PS-120-12V
» **Printer carriage**
» **Nails (3)**
» **Hex nuts (3)**

TOOLS
» **3D printer (optional)** Check out makezine.com/go/digifab-access to find a machine or service you can use.
» **High-speed rotary tool** e.g., Dremel
» **Locking pliers** e.g., Vise-Grips
» **Center punch**
» **Soldering station**

[+] Build tutorial and videos: instructables.com/id/OpenBraille

[+] 3D printing files: thingiverse.com/thing:2586738

[+] Project code: github.com/carloscamposalcocer/OpenBraille

[+] Follow: makezine.com/go/open-braille-facebook

Blind + Makers
THE BLIND ARDUINO PROJECT
ski.org/project/blind-arduino-project
blarbl.blogspot.com
Joint initiative by Smith-Kettlewell and Lighthouse Labs to help blind STEM students and electronics enthusiasts, offering hands-on workshops on Arduino, coding, and soldering.

BABAMM — BAY AREA BLIND ARDUINO MONTHLY MEETUP
meetup.com/Blind-Arduino-Monthly-Meetup
Regular meetup of visually impaired makers to share projects and information.

Braille and Beyond
ELIA FRAMES FONT
theeliaidea.com
ELIA Life Technology's new tactile reading system that builds on shape recognition and Roman characters to increase literacy in the blind community.

BLIND HELPER TOOLKIT
makerfairerome.eu/en/exhibitors/?ids=1762
Italian technical high school students designing electronics to assist the blind, including their own low-cost braille printer.

Promising Prototypes
ARDUINO SMART CANE
diyhacking.com/arduino-smart-cane-for-the-blind
Detects obstacles with an ultrasonic sensor and provides haptic feedback using a cellphone vibration motor.

COOKING OIL TEMPERATURE ALERT
makershare.com/projects/cooking-oil-temperature-detector
Arduino-based alarm to cue blind chefs when cooking oil is hot enough.

HOTGUARD WEARABLE HEAT DETECTOR
lepton.flir.com/community-showcase/hotguard
Wrist-worn gadget uses FLIR and ultrasonic sensors to detect hot stuff, prevent burns.

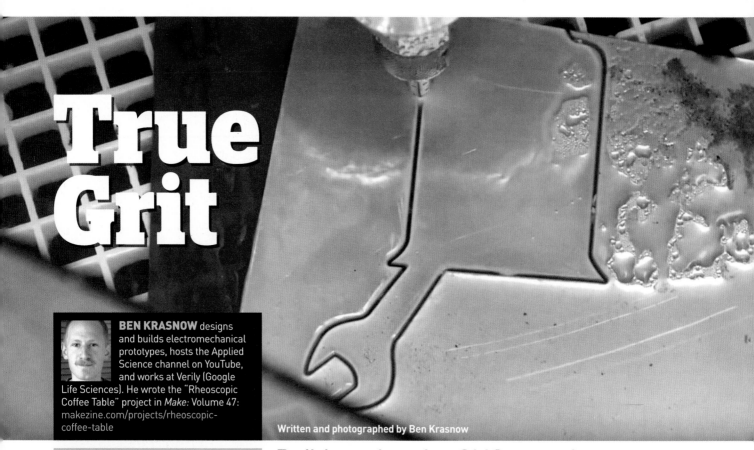

True Grit

BEN KRASNOW designs and builds electromechanical prototypes, hosts the Applied Science channel on YouTube, and works at Verily (Google Life Sciences). He wrote the "Rheoscopic Coffee Table" project in *Make:* Volume 47: makezine.com/projects/rheoscopic-coffee-table

Written and photographed by Ben Krasnow

TIME REQUIRED:
A Weekend

DIFFICULTY:
Intermediate

COST:
$800–$1,400

MATERIALS
- » **Pressure washer, gasoline powered, 3,100psi** Excell EPW2123100, Amazon #B01MZEIYM7
- » **Waterjet cutting head** Accustream #A2, accustream.com
- » **Waterjet orifice, 0.025"** Accustream #11007-025
- » **Waterjet nozzle, 0.045"** Accustream #12781-045-30, aka focusing tube
- » **High-pressure nipple and collar, ⅜"** Accustream #WJN6600600 and #13157-60-6
- » **Pressure washer hose coupling** Amazon #B01MXW4GG9
- » **Garnet abrasive, 80 grit** about $1/lb delivered from eBay
- » **Bucket, valve, and tubing** for flowing the garnet
- » **Bolt, stainless steel** to fit cutting head
- » **CNC router/engraver/mill** Find a Chinese 3040T on eBay under $500.

TOOLS
- » **Brazing or welding tools, wrenches, drill press**

Build an abrasive CNC waterjet cutter from a pressure washer for under $1,500

Waterjet cutters — everyone wants one but they're too big and expensive! So, for a long time I wanted to try building one out of a low-cost pressure washer. Last year I demonstrated halfway decent results on YouTube using an electric pressure washer, but the motor overheated and the pump developed leaks.

So I tried again with a stronger, gasoline-powered machine, and got excellent results that I'm confident you can replicate at home.

1. GAS PRESSURE WASHER
There are no modifications, it's completely stock (Figure Ⓐ). I take the hose right from the pressure washer and connect it to a waterjet nozzle system. I'm getting 3,100 psi at 2.8 gallons per minute.

What's doing the cutting in a waterjet cutter is the particles of garnet sand that are coming out at high speed. The faster they're going, the more efficiently they cut. That's why waterjet cutters use such high pressures — 30,000 to 90,000psi. Our rig generates much lower pressures, but we make it work by using a higher flow rate.

2. WATERJET CUTTING HEAD
Flow rate is controlled by the size of the tiny ruby orifice that admits water into the back of the waterjet cutting head (Figure Ⓐ). Luckily they're inexpensive, so you can try different sizes. I settled on 0.025" — relatively big, and still allows the pressure washer to operate normally.

To connect the pressure washer, get a pressure-washer hose coupling and then braze or weld in a nipple that fits into the back of the cutting head (Figure Ⓑ). You've got a lot of surface area to braze this together; it won't come apart at 3,000psi.

3. NOZZLE
The focusing tube is the cutting nozzle that comes out the bottom of the cutting head. It works sort of like a rifle barrel, where the high-pressure water accelerates the garnet as it moves down the tube, and shoots it out the end going really fast. Generally you want this nozzle to be 2.5 or 3 times bigger than the orifice. I bought the biggest one Accustream had, 0.045", so not even quite double, but the thing works!

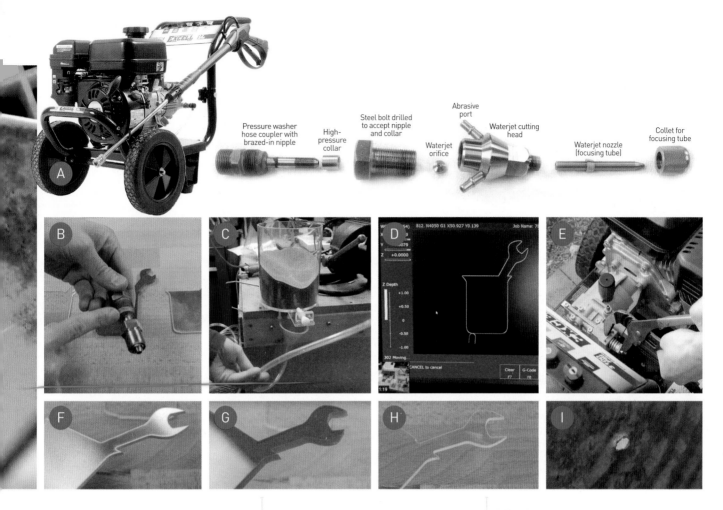

Pressure washer hose coupler with brazed-in nipple — High-pressure collar — Steel bolt drilled to accept nipple and collar — Waterjet orifice — Abrasive port — Waterjet cutting head — Waterjet nozzle (focusing tube) — Collet for focusing tube

4. GARNET ABRASIVE

The abrasive flow rate is critical. As you add more abrasive you get a faster cut rate, until the stream is completely saturated with garnet and can't do the momentum transfer in the focusing tube. I know that 0.4lb/min works fairly well — cuts ⅟₁₆" aluminum at 2" per minute — but I have a feeling you could turn this up quite a bit higher.

I used a simple gravity-feed setup (Figure C): bucket, valve, and tubing that connects to a port in the cutting head, where suction pulls the abrasive right in.

The abrasive costs about $1/lb, so about 20 cents per inch cutting ⅟₁₆" aluminum, and if you're cutting a thick material it's higher because your cutting speed is slower. Check how long your toolpaths are in Fusion 360 — for something intricate like a sprocket, all those little holes and teeth add up.

5. CNC MACHINE

There are plenty of small CNC machines on eBay that aren't particularly strong but are actually a great fit for waterjet cutting because the cutting forces are so low. You hardly need to clamp the workpiece at all.

All told, you're definitely under $1,500 to get a very serviceable waterjet cutter. And since the pressure washer is operating within its normal guidelines I expect a decent lifetime out of it.

OPERATION

Operation is pretty straightforward. Start the motor up and the pressure will rise. Squeeze the trigger to let the high-pressure water out of the nozzle. Then add the abrasive flow. Then start your CNC to move the nozzle over the workpiece (Figure D).

One problem you might run into is that the pressure washer has a built-in unloader valve: when you let off the trigger to stop the water flowing, this valve bypasses the output and lets the water recirculate. We're using an orifice a bit smaller than the one on the stock pressure washer wand, and so the unloader valve can sometimes cycle between bypass and flow. To fix that, try adjusting the valve for a slightly higher working pressure, or try a bigger orifice. I found that just squeezing the unloader valve slightly with channel-lock pliers would get the system flowing smoothly (Figure E).

Also, it gets messy. If you try to cut too quickly and the waterjet doesn't make it all the way through the material, then you get a big spray cloud. Better slow than sorry.

RESULTS

The precision is quite good — the jet is constant, so the real key to getting square, smooth cuts is how rigid your setup and how good your CNC is.

I estimated the kerf to be about 1mm and used Fusion 360's built-in CAM feature to generate the G-code.

F Aluminum — These edges are directly off the machine, I didn't smooth them at all.

G Acrylic — The cut quality is great, it's really 90° to the surface.

H Glass — The material everyone wants to waterjet. This also came out quite nice. You can see a little bit of frosting or fogging on the surface, from the abrasive getting thrown around. Cover your workpiece with tape to prevent this.

I Steel — I used a galvanized steel tub to hold all the water that comes out of the cutter. The machine was powerful enough to punch a hole in the tub from 1 foot away, very much to my surprise, and continued boring a hole into the concrete floor of my garage! ◆

[+] See the build and the DIY waterjet cutter in action: youtu.be/qAIDFaKhcZE

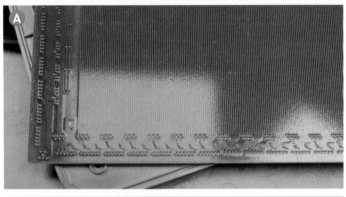

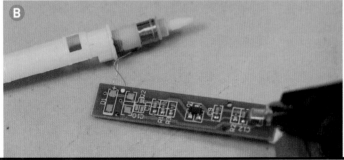

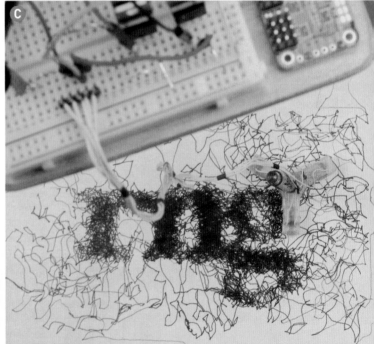

Tablet Teardown

Just what's inside those Wacoms and Huions, and how can you use them in projects? Written and photographed by Micah Scott

CAPACITIVE TOUCHSCREENS AND TOUCH SENSORS ARE UBIQUITOUS

on the computers around us now, and they're starting to be common on DIY projects as well. When you place your finger on a phone touchscreen, the conductivity of your fleshy bits causes a small change to the capacitive coupling between thin transparent electrodes printed under the glass. This is great for quick, coarse interactions with our pocket computers, but it isn't designed for accuracy, and the range is typically limited to direct contact.

For creative work, it's useful to have a precise input device that you can hold like a pencil. This isn't just a plastic stick, like you'd see on the resistive touchscreens from the Palm Pilot era, it's a thicker stylus that contains some active circuitry which forms a specific position reference that the tablet device can "see" from a distance. The pen typically also provides buttons and pressure sensors.

These *graphic tablets* or *pen tablets* have been popular with digital artists since the 1970s, but in the last decade or so prices have dropped enough to finally make them appealing for hardware experiments.

Wacom is the biggest name in tablet devices, and they have a unique (patented) method for powering the pen circuitry from the tablet, so you don't need to recharge or replace the pen batteries ever. The latest from Wacom will set you back hundreds of dollars, but the going rate for a used CTE-450 Bamboo Fun tablet, with an active area of 5.82"×3.64", is only $15.

We'll also look at an alternative made by Huion, popular with artists on a budget. The H610 is almost enough for a full page, with a 10"×6.25" active area. This device takes a simpler approach, with a battery-powered pen that transmits a continuous-wave signal back to the tablet. They're between $50 and $80 new.

HOW PEN TABLETS WORK

Pen tablets work by *mutual inductance*, as if the pen and tablet are each half of a transformer. This may sound familiar if you've read about RFID cards, but where a keycard reader only needs one coil, a tablet needs a whole array of coils to sense the pen's position (Figure Ⓐ). Just behind the pen nib you'll find a wire coil made from about 30 turns of thin copper (Figure Ⓑ). The tablet's counterpart is far simpler: just a few traces on one side of a circuit board.

This is the key engineering that made both tablets simple enough to build for a reasonable price. The coils are in fact implemented as two perpendicular low-resolution arrays, with horizontal coils on one side of the board and vertical on the other. The tablet's firmware can scan each axis separately to locate the strongest signal source, using the pattern of signal strengths among adjacent coils to estimate the true two-dimensional position of the pen.

With the constrained size of each coil, the tablet sees only a weak coupling to the pen's signal. The tablet uses a single chain of amplifiers to filter and boost this signal, shared among all the coils using a bank of analog multiplexer chips. There wouldn't be enough time to scan every coil and keep an interactive frame rate. Instead, the firmware needs to switch between a slower search pattern and a more targeted fine tracking pattern whenever the pen is located.

At this point the Wacom and Huion designs diverge. Huion's pen is a single-transistor oscillator. Pressure on the nib changes the frequency from 255 to 266kHz by tuning the inductor,

MICAH ELIZABETH SCOTT (scanlime) is a maker, reverse engineer, video producer, and live-streamer who likes to take things apart on camera and build complicated robots for her cat.

and the two buttons switch to 235 or 245kHz with additional capacitors.

The simplest Wacom pen would be a resonant LC circuit tuned to 750kHz. To transmit button and pressure status, an additional digital circuit modulates the resonant damping to send out individual bits of sensor data on each carrier burst.

For more details, see episodes 12 (youtu.be/nPab7pbOhBY) and 13 (youtu.be/j4AKwJERxOw) of my video blog (scanlime.org). Also see the journal PoC||GTFO, Volume 13:4 (archive.org/details/pocorgtfo13) for some extreme reverse engineering and repurposing of the CTE-450 tablet … to read RFID tags!

WHAT YOU CAN DO WITH THEM

There's plenty you can do with a graphic tablet without modifying it at all. You can emulate a Huion pen using a Raspberry Pi or Arduino with PWM. Emulate a Wacom pen by choosing a capacitor and inductor that resonate at 750kHz.

All modern tablets are USB input devices, but due to protocol complexities beyond our scope, it's helpful to have drivers that understand the compatibility modes used by individual tablet models. The easiest way to use a tablet in your project is to attach it to a Raspberry Pi, where these Wacom and Huion and many other tablets will be supported by the evdev interface in Linux.

You can start to think of a tablet as an absolute position sensor, like a two-dimensional pair of calipers. If you build a pen emulation into the build plate of your 3D printer, you could drive the x- and y-axes with DC motors and use the tablet for servo feedback. Or put a pen in the hand of an inaccurate robot arm, and use the tablet as fine position feedback to help it draw. Even a vibrating bristle bot can draw (Figure Ⓒ), with a little bit of position feedback from a pen tablet! ⊘

SHORT CUTS

Tips and tricks for getting the most out of your *miter saw*

Written by Jeremy S. Cook • Illustrated by Shing Yin Khor

POWER MITER SAWS **ARE SOMETIMES (MISTAKENLY) CALLED** *CHOP SAWS.* They work in a pretty similar way — you can use them to rend a long piece of material, like a 2×4 or PVC pipe, into smaller pieces — but miter saws have a few features that make them special. Most importantly, the saw blade is mounted on a pivoting mechanism, allowing you to make measured, angled cuts in your stock — *miter* cuts — rather than just 90° crosscuts. They also commonly have a sliding mechanism, similar to that of a radial-arm saw, that lets you make crosscuts on wider stock.

Compound miter saws can also pivot the blade off the vertical axis, allowing you to make *bevel* cuts. Combine these features, and you can make really precise, angled cuts with a single saw.

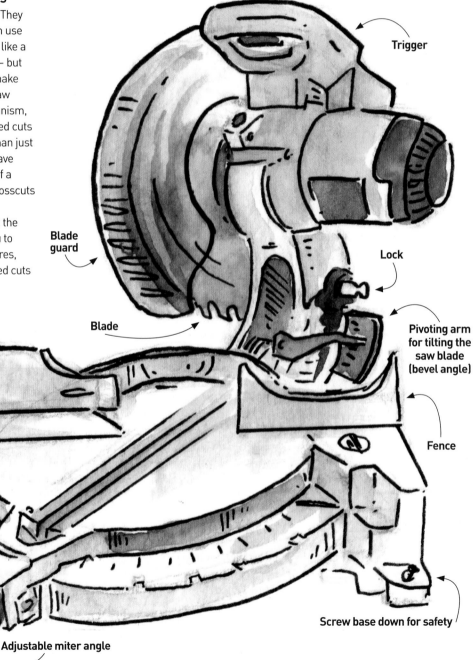

Trigger

Blade guard

Lock

Blade

Pivoting arm for tilting the saw blade (bevel angle)

Fence

Screw base down for safety

Adjustable miter angle

JEREMY S. COOK has a BSME from Clemson University, and worked for over 10 years in manufacturing automation. He writes for technical publications and pursues projects in electronics, robotics, CNC, woodworking, and more. Find his exploits on the Jeremy S. Cook YouTube channel, or on Twitter @JeremySCook.

Basic Use

Generally speaking, to cut something with a miter saw, you line it up against the *fence* (the vertical edge that a part naturally rests on while cutting), and pull the saw down on top of it with the trigger depressed. If your saw has a laser aiming device, this can be helpful, but keep in mind that you must accommodate for the cut width — the *kerf* — if the resulting piece is to be dimensionally accurate.

All miter saws have an adjustable mechanism that allows you to rotate the saw relative to the workpiece, giving a nice angled cut. Compound miter saws have a second rotational axis, allowing the blade to tilt relative to the floor. This allows for *compound* cuts that are angled relative to both the fence (miter) and the floor (bevel).

Take your time setting up your fence for the most accurate angles and cuts. Your fence is likely attached to the base of the saw with four bolts. Loosen all of them, then tighten the furthest left bolt until it's just barely snug, but you can still move the fence. With the blade locked down, use a speed square or machinist's square to ensure that the blade is at a perfect right angle to the fence. Then tighten the other bolts, making sure the blade is still square when you're done.

Tips

It's pretty easy to get a decent cut on a straight length of wood. Beyond that, review these tips:

» Wait several seconds for the saw to reach speed before beginning your cut. This can prevent splintered starts and rough cuts.
» For safety, let the saw stop before removing your piece.
» If you're cutting multiple pieces for a project where accuracy is secondary to speed, mark your fence with a felt-tip pen to indicate where the piece should end. This way you won't have to measure every time.
» Building a jig will improve the accuracy of repeated cuts. Screwing a block down where the board should start will give you the exact same cut every time. You can even screw directly into the saw base with self-tapping screws without changing the performance of the saw.
» If your material is not flat (say, PVC pipe with a fitting on one end), you may have to support the other end to keep your cut square.
» Clamp items that you're cutting if possible.
» When cutting thin material or short pieces, sandwich it between thicker sacrificial stock with toggle clamps.
» To shave a bit off the end, lower the non-spinning blade, and press the material snug to the stopped blade. Raise the blade, start the saw, and lower again. In effect, the blade is slightly "wider" when spinning than when stopped.
» When you're making wide cuts, raising your material can increase your cutting range by using the wider part of the blade.
» If your workpiece is too wide to cut in one pass, it's sometimes possible to cut and then flip and cut along the same axis.
» To avoid splintering, narrow boards should be cut with the best face up, and wide boards should be cut with the best face down.
» A shop vac can often be rigged to attach to the saw's dust chute.
» Always use eye and ear protection, and secure loose hair and clothing.
» When selecting a miter saw, take note of the sliding mechanism of the blade. Saws that slide on long rails may require a large amount of empty space behind the saw for its full range of motion. Some saws slide on a collapsing mechanism, which is ideal for compact workspaces. ●

Hep Svadja

STICK OUT
Create custom swag with a desktop cutter
Written and photographed by Hep Svadja

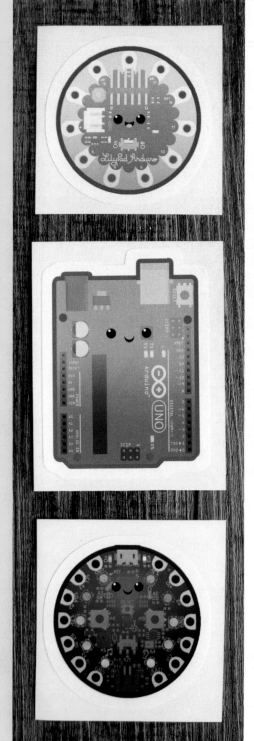

TIP: When creating your digital design, add a square cutting boundary, called a *weeding border*, around the entire complete design (shown above).

HEP SVADJA is *Make:*'s photographer and photo editor. In her spare time she is a space enthusiast, metal fabricator, and *Godzilla* fangirl.

DESKTOP CUTTERS, LIKE THOSE MADE BY CRICUT, SILHOUETTE, AND USCUTTER, ARE A POPULAR CRAFT ITEM. Pairing them with an inkjet or laser printer opens up a range of options like custom stickers, vinyl decals, and even heat-transfer vinyl designs for fabrics, to decorate and organize your life.

SOFTWARE

Some desktop cutters use proprietary software, while others let you import many file formats. Either way, it's important to understand the difference between vector and raster images. *Vector images* use paths and points to display and transfer your design. They can easily be sized up or down because the shapes are dynamically scaled. *Raster images* are made up of a set amount of pixels in predefined spaces, so they are not easily scaled up without losing detail.

Desktop cutting software tends to prefer vector files because they easily translate to cutting paths, but most software can also define cutting paths from raster images by tracing areas with great contrast either by hand or automatically.

VINYL DECALS

Cutting standard single-color decals is easy and fun, but *layering* vinyl decals allows a ton of creative possibilities and lets you use the many color options available, including glitter and holographic treatments.

TIP: If you're doing multiple layers, include *orientation marks* in your design, such as crop marks or top/bottom marks, to help you easily align and place each layer.

Cut your primary layer and use tweezers to pull out the excess material, or pin down delicate or detached areas (Figure **A**), a process called *weeding*. Apply *transfer tape* to the weeded design, and use a squeegee to firmly stick the transfer tape to the vinyl (Figure **B**). Lay down the first layer in the position you want the final design, then peel back the transfer paper and discard (Figure **C**). Repeat the process of weeding and applying transfer paper to the second layer. Position the second layer so that its orientation marks line up with those of the

Tote design and photo by Susy Zambrano – Instagram: @susy1017

primary layer (Figure **D**). Continue adding each successive layer, orienting to the marks (Figure **E**). When you apply the final layer, firmly adhere everything together and remove the orientation marks (Figure **F**).

HEAT TRANSFER

Heat Transfer Vinyl (HTV) is specifically created for fabrics, which makes it an excellent choice for decorating T-shirts and tote bags (Figure **G**) or adding cosplay details. Like decals, you can layer HTV to create complex designs, but keep in mind that the repeated pressings can cause some HTV materials to shrink — to minimize that, choose a high-heat tolerant HTV material for your base layer.

Use a flat heat press or flat household iron, and always use multiuse paper or a towel underneath for protection. Set each layer with a quick heat press of a couple seconds. After the final layer, press for half the recommended time to adhere all layers together.

STICKERS

Being able to use your own cutting paths lets you create custom packaging, planner stickers, and shaped stickers of all sizes (Figure **H**). Add a cutting layer when you design your printer artwork, or import your sticker designs into your cutting software and trace it to create the cutting paths.

Consider where you intend to place your stickers and what elements they may be exposed to when deciding what material to print on. Paper stickers are cheaper, but will fade in sun or moisture. Printable vinyl is stronger and can be laminated for waterproofing, but the inks can still fade. Using archival pigment-based inks may keep your stickers vibrant for longer — but it can cause some vinyl to shrink, so test it first.

Both inkjet and laser printers will print stickers well. Inkjet tends to be better for color and gradients, and while cartridges are expensive, you can refill them yourself. Laser printers are better for precision line work and graphics, and even though the

upfront cost can be more, over time you will pay less in toner. Regardless of which you choose, make sure you always use paper designated for your printer type.

CAUTION: Don't use inkjet paper in a laser printer. This can cause the vinyl sticker material to melt and ruin both your design and your machine.

MORE IDEAS

Visit *Make:* online to learn how to use your vinyl cutter for making **silk screens** (makezine.com/projects/vinyl-silk-screen-printing), **spray paint stencils** (makezine.com/2017/06/07/digital-stencil-design), and even **custom circuit boards** (makezine.com/projects/vinyl-cut-pcb-resist).

No matter what you decide to make, most manufacturers have online guides for getting the most out of your cutter, and you can also seek help from the many active online communities. ⊘

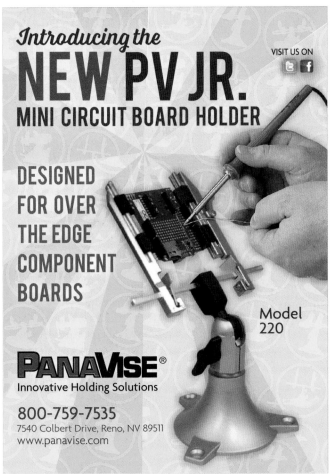

TOOLBOX
GADGETS AND GEAR FOR MAKERS
Tell us about your faves: *editor@makezine.com*

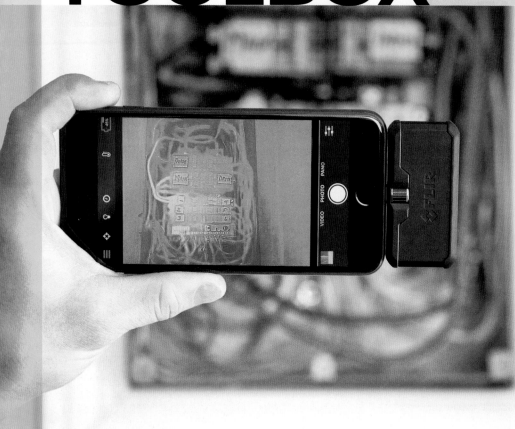

FLIR ONE PRO
$400 flir.com

The FLIR One Pro is a thermal imaging camera that works in cooperation with your smartphone. It uses the processor and screen from your phone and can hybridize data from the thermal sensor and your phone's built-in camera. The Pro model supports temperatures up to 400°C, which is plenty for watching how a 3D printer bed heats up, or how a loaded-down CPU chip on a single board computer handles different cooling mechanisms.

While still not a low cost solution, if you are working on a project where heating or cooling are important, the FLIR One Pro can save you hours of time over probing around with a thermometer. –*Matt Stultz*

WORK SHARP KNIFE SHARPENERS
$60 (Combo) – $150 (Ken Onion Edition) worksharptools.com

All blades eventually dull, even the fancy German chef's knife I treated myself to years ago. Work Sharp's motorized, belt-focused sharpeners caught my eye due to their built-in angle guide and the company's claims that the resulting convex blade edge stays sharp longer than flat-angled results. I tried their basic Combo model and the top-of-the-line Ken Onion model on a couple of my blades and quickly got the results I wanted. So quick that I ran over to my in-laws' house to sharpen all their knives too.

The Combo, with a set 25° angle and a fine-grit belt, is meant for the casual knife owner who wants to maintain an edge on decently sharp knives. But for badly worn or damaged blades, or different angles, the larger Ken Onion model will let you change belts and settings to get a razor-sharp edge on just about anything. Be careful though, because it can also chew through a good blade if misused — practicing on a cheap knife is recommended. –*Mike Senese*

PI-TOP $320 pi-top.com

There are plenty of build-your-own Raspberry Pi laptop kits out there, but the modular Pi-Top is designed with the maker in mind. The Pi itself is easily exposed, giving access to a prototyping space with a breadboard and components for three included electronics projects. More electronics peripherals are also available to further customize your machine. The Raspberry Pi comes preloaded with pi-topOS and includes CEEDuniverse, a coding-based role playing game that merges with the pi-topCODER programming system for a well-rounded learning environment. The Pi-Top integrates well into educational programs, with curriculum design tools and lesson plans, and the 8-hour battery life means it lasts the whole school day. But this laptop isn't just for kids — I use mine in the lab for rapid prototyping builds without hauling around an HDMI monitor. –Hep Svadja

BRICK'R'KNOWLEDGE ARDUINO CODING SET

$245 brickrknowledge.de/en/sets/engineering/arduino-coding-set

I've tried lots of systems over the years that make it "easy" to plug sensors and outputs into an Arduino to quickly mock up a circuit. Magnet-based solutions tend to disconnect easily or have bad circuits, and wire and socket connections are often difficult to plug and unplug. The Arduino Coding Set from Brick'R'knowledge solves these problems with large, easy-to-connect bricks.

The kit contains 45 blocks with basic items like LEDs and pushbuttons, along with fun components like a relay switch and an OLED display. The core of the system is an Arduino Nano, which is easy to upgrade if a new, more desirable model comes out, or if you accidentally burn out the board. It's a great kit for getting started with Arduino. –Matt Stultz

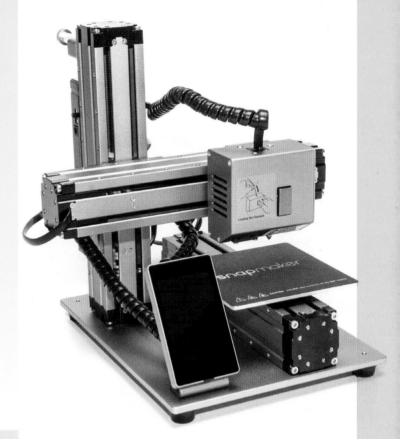

SNAPMAKER

$800 snapmaker.com

The Snapmaker is a hybrid 3D printer, laser etcher, and CNC cutter all in one machine. This isn't the first hybrid we've tested, and while hybrids are capable of doing all the tasks they advertise, they tend to do so with compromises. But with the Snapmaker, I still found it to be a great little printer.

The Snapmaker comes as a kit with all the tools you need to assemble it (a couple Allen wrenches), and I had it up and running in about 45 minutes. This is a small machine with only a 125mm×125mm×125mm build area for 3D printing. There's also a bright, full color LCD that feels more like a cellphone than part of the printer, and everything is connected with cables that snap into place. The structure could be a lot stiffer for CNC milling, but it works fine for laser etching and 3D printing.

To change out tools you simply remove a couple screws, unhook the control wire from the old toolhead, screw in and replace the wire for the new toolhead, and change out the base when not 3D printing.

If you have the room and money for a laser cutter, CNC machine, and 3D printer, buy them separately, but if space is a premium, the Snapmaker might be the machine for you.

–Matt Stultz

SHOW & TELL

Dazzling projects from inventive makers like you

Sharing what you've made is half the joy of making. Want to be featured here? Showcase your project on **makershare.com** or tag us on Instagram with **#makemagazine**.

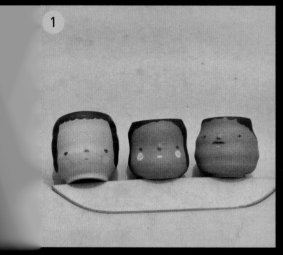
1

2

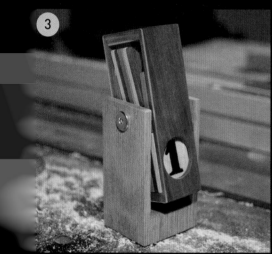
3

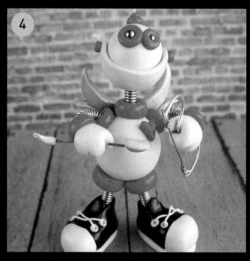
4

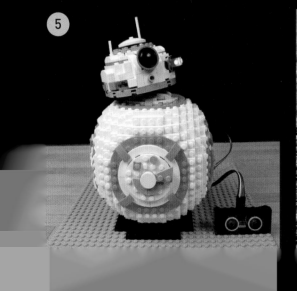
5

6

1 **Hana Ward** co-founded Uno+Ichi as a small indie ceramics studio. Though she now operates solo, her playful mugs, cups, and vases — with stoic faces and whimsical names — are sold around the globe. uno-ichi.com

2 Remember that incredible podracing level in *Lego Star Wars: The Video Game*? **Alexis Dos Santos** built a real-life equivalent that's just as fun to play. flickr.com/people/tkel86

3 **Eric M. Thompson** carved and assembled this perpetual flip calendar on his YouTube channel, *Measured Workshop*. Change the date with one easy flip! youtube.com/channel/ UCvQcbUdIMoDc8tOG0drAupg

4 **HerArtSheLoves** sculpts custom robots. Each bot has its own colorful personality, ranging from adorably innocent to playfully sassy. theawesomerobots.com

5 These Lego *Star Wars* sets have been redesigned to move, light up, and make sounds whenever they detect motion. So far, **Joshua Zimmerman** has hacked an X-wing, BB-8, and ARC-170 starfighter. makershare.com/portfolio/ joshua-zimmerman

6 Inspired by Studio Ghibli, **Winnie Lavinthal**'s handmade terrariums feature cute renditions of some of the creatures from the films. A few even glow in the dark! etsy.com/ people/Choconup543